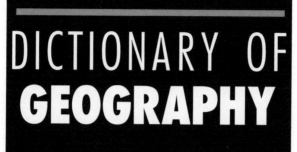

DICTIONARY OF
GEOGRAPHY

**MALCOLM SKINNER,
DAVID REDFERN & GEOFF FARMER**

Series editor Ian Marcousé

FITZROY DEARBORN PUBLISHERS
LONDON · CHICAGO

Copyright © 1996, 1999 Malcolm Skinner, David Redfern and Geoff Farmer

This edition based on *The Complete A-Z Geography Handbook*, first published in the United Kingdom by Hodder and Stoughton Educational, 1996

Published in the United States of America by
Fitzroy Dearborn Publishers
919 North Michigan Avenue
Chicago, Illinois 60611
USA

A Cataloging-in-Publication Record is available from the Library of Congress

ISBN 1-57958-154-4

First published in the USA 1999

Typeset by Wearset, Boldon, Tyne and Wear, England and Alacrity, Banwell Castle, Weston-super-Mare, England

Printed in the UK by Antony Rowe Ltd., Chippenham, UK

PREFACE

The *Dictionary of Geography* is an alphabetical text designed for ease of use. Each entry begins with a precise one-sentence definition and is then developed in line with the relative importance of the concept covered, often through the use of worked examples and illustrations. All entries are carefully cross-referenced through the use of italics.

We hope that the *Dictionary of Geography* proves an invaluable resource for users.

Malcolm Skinner, David Redfern, Geoff Farmer

ACKNOWLEDGMENTS

The authors are grateful to the individuals who have provided the patience and additional support for this project. Family members are first and foremost for these thanks – Judith and Katharine Skinner, Tina Redfern and Angela, Andrew and Clare Farmer. Other people to be thanked are the team at Hodder & Stoughton, London, especially Tim Gregson-Williams and Julie Hill, and the series editor Ian Marcousé. Perhaps we should also include a special thanks to our long-suffering geography colleagues at Bury Grammar School, Adwick School and Winstanley College.

Researching, writing and editing a book of this size requires teamwork and above all hard work. A considerable amount of time went into checking the entries, but if any mistakes have slipped through, the authors accept full responsibility. Geography covers a huge range of topics, many of which change through time and space. The concepts of the subject are constantly being updated and revised, and it is inevitable that some of the ideas contained herein may already be outdated, or subject to review. This does not, however, alter the relevance or usefulness of geographical knowledge.

ablation is the collective loss of water from an ice sheet or glacier. This loss can take a variety of forms:

- melting on the surface, melting internally, melting at the base or melting from the ice front
- *calving* of blocks of ice from the front of the ice mass into water to create *icebergs*
- *evaporation* of surface snow
- the process of *sublimation*
- the blowing away of snow by strong winds.

Ablation is dominant in the lower parts of a glacier or ice sheet, and takes place more readily during warmer times.

abrasion is the scraping, scouring, rubbing and grinding action of materials being carried along by moving natural features such as rivers, glaciers, waves and strong winds. Rivers either carry rock fragments in the flow of the water, or drag them along the river bed, and in doing so wear away at the banks and bed of the river channel. Similarly, glaciers use rock fragments embedded in their base; waves hurl pebbles and sand grains at a cliff face; and strong winds use rock fragments in a sand-blasting manner. In each case, the rate at which abrasion takes place depends on the amount of material being carried, and on the relative resistance of the substances being eroded. Abrasion is very effective on less resistant surfaces, but is less so on harder surfaces.

abyssal plains are the relatively level areas of the ocean floor that can be found at approximately 5000 meters below sea level. The North Atlantic abyssal plains, for example, lie at a depth of about 5500 meters.

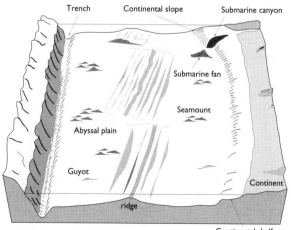

Major submarine features

Occasionally volcanic peaks (seamounts) rise steeply from the plain, sometimes reaching the surface as islands. Detailed surveying of the Pacific Ocean has revealed the presence of numerous flat-topped mountains (*guyots*) rising to within a kilometer of the surface.

accessibility is a measure of the ease with which one can reach features in the wider environment. This may be accessibility to:

- a physical feature – a river, a coastline, an area of attractive scenery
- a built feature – a freeway, a city center
- a human feature – a place of employment, education or leisure.

Accessibility is not simply distance – it may involve time and cost, or other factors such as the availability of services and the ability to qualify for those services.

In transportation studies, accessibility is a measure of the ease with which one can travel from one part of a *network* to another part of the same network. Indicators such as the *associated number* are used to measure how accessible a place is in a network.

accretion is the growth of a natural feature by enlargement. For example, hailstones grow by accretion. Droplets of water are carried upward in a cumulonimbus cloud by strong vertical air currents and freeze. These ice crystals then fall from the higher parts of the cloud, and as they fall they enlarge by colliding with more supercooled water droplets (water droplets existing at temperatures below freezing), which freeze on contact. Subsequent updrafts of air carry the enlarged hailstone upward again for the cycle to be repeated. In this way a hailstone grows by a series of concentric layers.

accumulation is the net gain in an ice mass. This gain can take place in a number of ways:

- by the precipitation of snow onto the surface of the ice mass
- by the refreezing of meltwater
- by snow avalanching from the surrounding slopes on to the ice sheet/glacier
- by strong winds drifting snow from other areas.

Accumulation is dominant in the upper parts of a glacier or ice sheet, and takes place more readily during colder times.

acidification is the consequent effect of *acid rain* falling on an area. Rain is naturally acidic (a weak carbonic acid), but its acidity is increased by atmospheric pollution.

By draining through soils, acidified rainwater leaches out the soil bases, such as calcium, magnesium and potassium, and replaces them with hydrogen. Aluminum is mobilized, and carried by throughflow to accumulate and become concentrated in lakes, where it poisons fish stocks. High concentrations of aluminum in a soil also poison tree roots. Underground water supplies also become more acidic as a result of passing through such soils.

Acidification of soils may also be caused naturally in areas based upon nonlime bedrocks, such as granite.

acid lava contains a high proportion of silica. These lavas have a high melting point, are very viscous, solidify quickly and so do not flow very far. Acid lavas build high, steep-sided volcanic cones. They may also solidify in the vent of the cone and cause recurrent and explosive eruptions.

acid rain (or acid deposition) consists of the dry deposition of sulfur dioxide, nitrogen oxides and nitric acid, and the wet deposition of sulfuric acid, nitric acid and compounds of ammonium from *precipitation*, mist and clouds.

Acid rain leads to direct damage to trees, particularly *coniferous* trees. It produces a yellowing of the needles and strange branching patterns. It also leads to the *leaching* of toxic metals (aluminum) from soils, and to their accumulation in rivers and lakes, which in turn causes the death of fish.

Acid rain is also blamed for damage to buildings, particularly those built of limestone, and to health problems in people such as bronchitis and other respiratory complaints.

The major causes of acid rain are the burning of fossil fuels in power stations, the smelting of metals in older industrial plants, and exhaust fumes from motor vehicles.

Various solutions to acid rain have been suggested:

- using catalytic converters on cars to reduce the amount of nitrogen oxides
- burning fossil fuels with a lower sulfur content
- replacing coal-fired power stations with nuclear power stations
- using flue gas desulfurization schemes, and other methods of removing sulfur either before coal is burned or after
- reducing the overall demand for electricity and car travel.

active layer is the name given to the upper layer of soil that undergoes seasonal thawing in *periglacial* areas that experience *permafrost* conditions. As summer temperatures rise thawing takes place from the surface downward, leaving a permanently frozen zone at depth in the soil. The thickness of the layer depends on local conditions but may extend to 5 or 6 meters. As the ice in the active layer melts, large volumes of water are released; this water is unable to drain through the lower layers of the soil because the permafrost creates an *impermeable* lower boundary. The top layer may become saturated, which reduces the friction between soil particles, and as a result, even on slopes as low as 2°, the active layer can move downslope. Such movements, which are given the general name *solifluction*, can produce features such as solifluction lobes and terraces. Other periglacial processes such as frost heaving, ice wedging and patterned ground take place in this active layer.

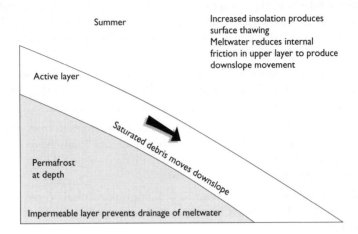

Summer

Increased insolation produces surface thawing
Meltwater reduces internal friction in upper layer to produce downslope movement

Active layer

Saturated debris moves downslope

Permafrost at depth

Impermeable layer prevents drainage of meltwater

The active layer

adiabatic temperature changes take place in rising (or sinking) air; these *lapse rates* are independent of the conditions in the surrounding environmental air. Adiabatic lapse rates refer to the temperature change for a parcel of air that has no exchange of heat or moisture with the surrounding environmental air; as it rises it will cool at a rate that is predictable and independent of environmental temperature. Conversely, sinking air will warm up according to adiabatic laws. The *Dry Adiabatic Lapse Rate (DALR)*, which applies to air that is not saturated (i.e. *relative humidity* is less than 100%), is 1°C per 100 meters. Saturated air has a lower lapse rate, which varies according to the temperature of the air. As the air reaches 100% humidity condensation of water releases latent heat, which offsets the normal decrease of temperature with height. The *Saturated Adiabatic Lapse Rate (SALR)* is normally taken to be 0.5°C per 100 meters.

adret is the slope of a hill or valley that faces toward the sun. In the northern hemisphere they face south or southwest and therefore are warmer because they receive more *insolation*.

adsorption is the physical or chemical bonding of liquids or gases to solid particles. In the clay-*humus* complex of a soil nutrients can be held in combination with humus substances and attached to the small clay particles. The clays have a negative electrical charge at their surface and to balance this a layer of positive ions (cations) coats the clay particles. These ions are adsorbed (stuck) to the clay.

Advanced Light Water Reactor program is the name given to research attempting to design water-cooled nuclear-powered reactors that are safer, simpler and cheaper than existing models. The program has produced designs for reactors with cooling systems that will continue to operate through major accidents.

advection is the horizontal transfer of heat. When a warm, moist air mass moves over a cooler land or sea surface it is cooled from below. If the temperature of the air mass cools below *dew point* an advection fog is formed.

aeolian processes involve the erosion, transportation and deposition of particles by wind action and are particularly important in desert areas where there is insufficient vegetation and ground moisture to bind the soil together. There are two main processes of wind erosion: *deflation* and *abrasion*. Wind can move small particles in suspension and larger grained material by *saltation* or surface creep.

Aeolian features include *deflation* hollows, rock pedestals, *yardangs*, *zeugens* and sand *dunes*.

aerial photographs are taken from above an area, looking down. Some aerial photographs have been taken with the camera pointing vertically down. Such photographs are very similar to large-scale maps, with a constant scale over the whole area shown. Their main weakness is that they present a level surface, and relief features are difficult to see.

An oblique aerial photograph is when the camera has been pointing at an angle to the ground. The scale on such photographs is more variable. The foreground shows a smaller distance, whereas the areas further away at the top of the photograph have a greater "real" distance between them. Heights and shapes of features and buildings are more easily seen.

afforestation is the deliberate planting of trees, usually where none have grown previously. Where planting takes place on land that was formerly wooded, but has been cleared, it is referred to as reafforestation.

African Alternative Framework: a blueprint put forward at a conference of African states in July 1989 for an "economically sound and socially just Africa." This was a reaction to the belief that many policies on Africa were "overwhelmingly inadequate." The Framework suggested that:

- for economic reform programs to succeed they should be seen to be suggested by Africans themselves
- the diversity of African situations cannot be dealt with by a standard formula for all
- improvement in the very low levels of health, nutrition, education and productivity cannot be put to one side while resources are devoted to debt-service payments.

agglomeration is when several firms choose the same area as their location in order to minimize costs. This can be achieved by *linkages* between firms and their supporting services. Firms are also able to achieve *economies of scale* with such concentrations. Agglomeration was one of the factors that Alfred *Weber* suggested could affect the location of industry.

aggradation is the building up of deposits by rivers within channels when they are forced to drop the material (*load*) they are carrying. This results in bed elevation.

agribusiness is the name for farming *systems* that are increasingly organized around scientific and business principles. Such businesses are linked firstly to agricultural supply industries upstream from the farm, which provide inputs such as chemicals, feedstuffs and machinery; secondly to the food processing industries downstream of the farm. These systems therefore exhibit *vertical integration* and are involved with most stages on the *agricultural chain*. The companies concerned are often *transnationals* such as Unilever and Nestlé, who in many cases seek out enterprising farmers with whom they can place contracts.

agricultural chain: at its simplest level this is a sequence of events linking the cultivation of crops or the tending of livestock to the food that is consumed on the table. In *subsistence* economies the chain can be relatively simple as the farmer's family eats all that he or she produces. On the other hand the chain can be extremely complex, especially when *agribusinesses* are concerned. Such chains include growing, processing, collection, packaging and distribution.

agrochemicals is the term describing the chemical inputs into farming that are used in order to increase productivity through:

- *fertilization* of the soil
- the control of pests and diseases by *pesticides, herbicides* and *fungicides.*

Several problems have occurred with the use of chemicals in farming:

- contamination of water courses and water supplies (*eutrophication*)
- growth of *algal blooms*
- damage caused to wildlife, some of which can be beneficial to the farmer
- poisoning of people through indiscriminate or accidental spraying
- concentration of *pesticides,* etc. in organic tissues in the *food chain* endangering aquatic life and human health
- possible links to diseases such as cancer
- rise of resistant wildlife or "super-pests."

aid is the giving of resources by one (donor) country, or organization, to another poorer (recipient) country with the main aim of improving the economy and/or living standards in the latter country. The resources given may be in the form of money, food, goods, technology, education or skilled people. There are three main types of aid:

- bilateral aid – aid from one country to another country. Often such aid is only granted on certain terms. The donor country imposes conditions on the recipient country such as contracts for buildings and projects, and preferential trade links
- multilateral aid – richer countries give money to an organization (e.g. *World Bank, International Monetary Fund, European Union*), which then redistributes the money to poorer countries. These organizations may withhold aid if they disagree with the recipient country's economic and/or political system
- voluntary aid – organizations such as Oxfam and March of Dimes collect money from the general public and then spend it on specific, usually small-scale projects in poorer countries.

AIDS is the abbreviation for Acquired Immune Deficiency Syndrome. It is the result of a viral infection called HIV (Human Immunodeficiency Virus), which causes a breakdown in the immune system of the human body. It is transmitted through the exchange of body fluids, notably blood, semen and vaginal fluid. It is associated with sexual activity and drug abuse, but has also been transmitted via blood transfusions.

The world distribution of AIDS is:

- those countries where the disease is mainly associated with drug abusers, homosexuals and bisexuals – North America, Western Europe, Australia
- those countries where the disease is widespread among the heterosexual population, and where vertical transmission (mother to child) is common – African countries south of the Sahara
- those countries where the disease has been introduced by travelers from the other two types of country.

air mass: an extensive body of air in which there is only a gradual horizontal change in temperature and humidity at a given height. *Lapse rates* are also almost uniform. These properties result from the air stagnating over a particular part of the Earth's surface for several days or even weeks. They are usually associated with large stationary *anticyclones*, often over ocean surfaces, deserts, ice-covered areas and large plains.

albedo is the amount of incoming solar radiation (*insolation*) that is reflected by the Earth's surface and the atmosphere. The planetary average for the albedo is 32% of insolation, but this amount varies considerably from place to place. Dark-colored areas of the world, such as coniferous forests, reflect small amounts of insolation (10%), whereas light-colored areas, such as deserts, reflect larger amounts (35%). Fresh snow and ice have very high albedos, up to 90%. Water surfaces also have variable rates, but here the angle of incidence of the sun is also important. Still water surfaces, with a vertical sun, reflect low amounts (3%), yet a low angle of incidence (15°) results in a high albedo (50%). Disturbed water surfaces have low albedos.

alcohol fuels are created by the fermentation of crops such as sugar cane, cassava and maize to produce ethanol. Ethanol has been used as an alternative to gasoline in motor vehicles in Brazil, such that 75% of Brazilian cars now run on ethanol. The main advantages that have arisen are its cheapness relative to imported oil, employment creation in sugar plantations and reduced levels of air pollution. (See also *biofuels*.)

algal blooms are the excessive growth of *phytoplankton* in water bodies, many of them producing toxins that are harmful to wildlife. Such toxins are known to affect humans, and in some cases fatalities have been reported. The increased occurrence of these blooms is often the result of *pollution* through sewage and agricultural run-off containing high levels of nitrogenous *fertilizer*.

alluvial fans are cones of sediment deposited by rivers emerging from steep-sided valleys in an upland area onto a lowland plain. At such points the rivers spread out, decreasing their velocity, dissipating energy and therefore depositing sediment. Fans derived from less resistant rocks are larger than those derived from more resistant rocks. Consequently, their size is variable, ranging from 100 meters to several kilometers across. In arid areas alluvial fans tend to be deposited by *mudflows*. Because of the variability of these flows, and the lack of vegetation cover, they are subject to frequent changes in character, and may be deeply dissected and unstable.

alluvium is the sediment deposited by a river. It can be deposited in two ways:

- when a river overflows its banks during times of flood, clays, silts and fine sands are deposited across the *floodplain*
- through the migration of a river *meander*, which leaves behind the remains of the *point bar* created on the inside bank of the meander. Such deposits consist of sands and gravels.

alternative energy sources are those that do not rely on the burning of fossil fuels. There are a number of such sources: *tidal energy*, solar energy, *wind power, geo-thermal energy*, wave power, and various forms of *biofuel*. Some of these energy sources are *renewable*, whereas others are not.

alternative technology is a feature of *economically less developed countries* in which hi-tech industries are both expensive and inappropriate for the needs of the people and the environment in which they live. Projects tend to be:

- labor-intensive because of the high rates of unemployment
- low-cost, using technologies based on local resources and skills
- in harmony with the local environment creating a more sustainable way of life.

(See also *intermediate technology*.)

ALWR program: see *Advanced Light Water Reactor program*.

anabatic is the name given to the motion of air up slopes resulting from *convection*. In upland areas by day, heated air will move toward the head of the valley and also up the valley sides. This is also known as a valley wind.

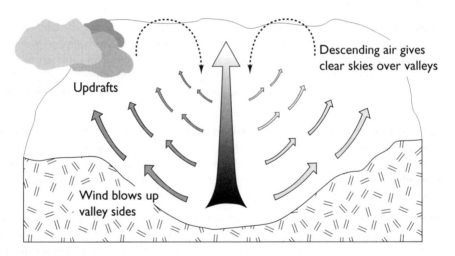

Anabatic flow

anaerobic refers to conditions within the soil where free oxygen is not present. Such conditions are usually the result of waterlogging where the pore spaces are filled with stagnant water that becomes deoxygenized. The reddish-colored ferric iron compounds of the soil are then chemically reduced to the gray-blue ferrous compounds. Under such conditions the rate of decomposition of organic matter will be slowed down, which leads ultimately to the formation of *peat*.

antecedent drainage is a form of discordant drainage that occurs when an ancient river is able to maintain its course across more recently uplifted fold structures. The rate of downcutting by the river must be equal to, or greater than, the rate of uplift. If the upfolding is more rapid, but not fast enough to divert the river, a convexity may develop in the long profile of the river where it crosses the fold system.

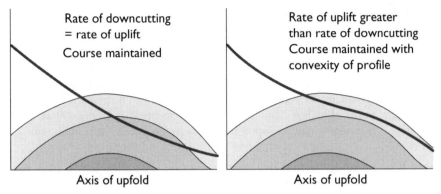

Rate of downcutting = rate of uplift
Course maintained
Axis of upfold

Rate of uplift greater than rate of downcutting
Course maintained with convexity of profile
Axis of upfold

Antecedent drainage

anticyclone is an area of high atmospheric pressure that is usually slow moving or stationary. Anticyclones are generally larger than depressions (up to 3000 km across) and are dominated by subsiding air, which produces warming and a decrease in *relative humidity*. In the summer they bring hot, sunny conditions with little cloud or rain, although clear skies at night can lead to *inversions*, which produce dew, mist and coastal fogs. In the winter these stable conditions favor the development of fog and frost, and pollution may be trapped in the lower layers of the atmosphere by the inversion.

anvil clouds are associated with *thunderstorms*. The rapidly rising air in a thunderstorm is forced to spread out sideways when reaching the *tropopause* and consequently the clouds produced by the rising air create a characteristic anvil shape.

apartheid was the policy of separate development operated in South Africa from 1948 until the early 1990s. It involved planned racial *segregation* and banned the mixing of races through marriage. The population was divided into three groups; the whites who were regarded as "first class" and had full political, social and economic rights; the coloreds and Indians who enjoyed some rights but were seen as "second class"; and blacks who had virtually no rights outside the "homelands" where they were forced to live. Apartheid, which means "living apart," established segregation in housing, the blacks being forced to live in townships such as Soweto (Johannesburg) and Crossroads (Cape Town), which were well away from white residential areas. These settlements lacked services and infrastructure and were usually surrounded by large squatter camps. Schools and universities were segregated by law and whites had their own buses and sections on beaches, in restaurants and in places of entertainment. To achieve residential segregation, African homelands (Bantustans) were established in rural areas. In these areas the blacks were allowed some of the political rights that they were not able to exercise in the rest of the country.

appropriate technology is a term used in relation to the supply of aid when the technical expertise and equipment provided by the donor country or organization are properly suited to the conditions that exist in the receiving nation. Rather than short-term "stop-gap" measures, the application of appropriate technology may lead to *sustainable development*. In countries experiencing food shortages it is more appropriate to improve existing farming methods and equipment rather than attempt to introduce totally different production methods.

aquaculture is the management of water environments, other than the sea, for the purpose of increasing production and controlled harvesting of plant and animal food sources, for example fish farming in man-made tanks, pools, rivers and lakes.

aquifer is the name given to a *permeable* rock that can store and transmit water. The accumulation of water is most effective when rocks such as limestone, sandstone or chalk are underlain by an *impermeable* rock such as clay. The water can percolate through the permeable layer but its downward movement is prevented by the impermeable boundary. When tapped by wells aquifers are an important source of water supply.

arable is a type of agriculture that concentrates on the cultivation of plant crops such as grasses, cereals, vegetables, root crops and animal foodstuffs.

arête: the name given to a narrow, knife-edged ridge with steep sides found in upland glaciated regions. They result from the formation of *cirques (corries)*, the arête representing the area remaining between two of these features after they have been enlarged through glacial erosion and *freeze-thaw action.* ·

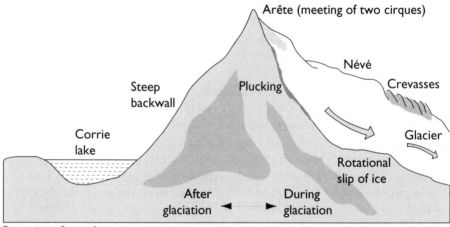

Formation of an arête

arid/aridity: areas where the climate is extremely dry. Such climates were defined as being areas with less than 250 mm of precipitation per year but in reality, this definition is far too simple. Several climatologists have attempted to devise a quantitative index expressing the relationship between *precipitation* and *evapotranspiration* that determines aridity. In those areas of the world where there is little precipitation annually or where there is seasonal *drought* the calculation of *potential evapotranspiration* is used. The best known aridity index was put forward in 1931 by Charles Warren Thornthwaite.

Thornthwaite's Aridity Index:

$$\text{Moisture Index (Im)} = \frac{100 \times \text{water surplus} - 60 \times \text{water deficit}}{\text{PE (Potential evapotranspiration)}}$$

Definition of climates   Im   0 to −20   dry subhumid

                                         −20 to −40   semiarid

                                         under −40   arid.

arithmetic average (mean): the arithmetic mean or average of a distribution is a *measure of central tendency*, which is calculated by dividing the total value by the number of occurrences.

FORMULA $\dfrac{\text{sum of values of the variables}}{\text{sum of numbers in the set}}$

The mean is a useful measure and can be subject to further calculations but it is distorted by extreme values and may not be a whole number or the same as one of the items in the distribution.

Asian Tigers: the name given to four countries of the *Pacific Rim* that have experienced phenomenal economic growth in the last 25 years. The countries are Hong Kong, Singapore, South Korea and Taiwan. Much of the growth has been based on the development of *manufacturing industry*.

aspect is the direction toward which a slope faces. In the northern hemisphere, north-facing slopes (*ubac*) receive less *insolation* than south-facing slopes (*adret*) and are therefore generally cooler.

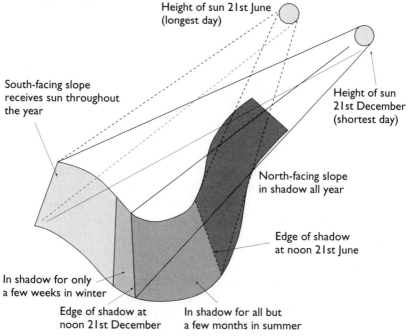

Sunlight received in a northern hemisphere alpine valley (orientated east–west)

associated number: a measure of *accessibility* in a *network*. It is the minimum number of edges taken to reach the furthest other vertex from the vertex being considered. Lower numbers indicate greater accessibility.

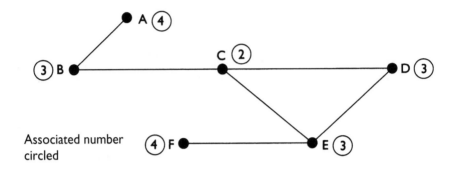

Associated number

asthenosphere is a layer in the Earth's *mantle* in which the rocks are soft and easily deformed. It is several hundred kilometers thick and lies below the *lithosphere*, which has a higher viscosity and is much more resistant to deformation. Temperatures increase with depth in the mantle until, at about 80 km, they reach 1400°C. Although different minerals have slightly different melting points this temperature is very close to the melting point of the mantle rocks and therefore these rocks will soften. The temperatures in the asthenosphere are high enough to cause deformation. There is a sharp boundary between the two zones at about 80 km depth, but as this can vary it is more accurate to describe the boundary as occurring at the 1400°C isotherm.

atlantic type coastline: occurs when tectonic processes have created folding of rock strata that is at right angles to the trend of the coastline. A series of bays and headlands develops as in the coastline of southwest Ireland.

atmosphere is a mixture of transparent gases held to the Earth by gravitational force. It consists of mainly nitrogen (78.09%) and oxygen (20.95%) by volume. Other gases include argon, carbon dioxide and traces of hydrogen, neon, helium, krypton, xenon, ozone, methane and radon. By international convention the upper limit of the atmosphere is assumed to be 1000 km, but, due to gravity and compression, most of the atmosphere is concentrated near to the Earth's surface. About 50% lies within 5.6 km of the surface and 99% within 40 km. Most of our climate and weather processes operate within 16–17 km of the surface in the zone of the lower atmosphere known as the *troposphere*. In this layer temperatures generally decrease with height (averaging 6.5°C per km). The top of this layer is marked by the *tropopause* where temperature remains fairly constant. This layer, which can act as a *temperature inversion*, forms an effective ceiling to any *convection* in the troposphere and an upper limit to weather systems. Carbon dioxide absorbs long-wave radiation from the Earth and is important in plant photosynthesis. Ozone absorbs and filters out ultraviolet radiation leading to a warming of the upper layer of the atmosphere, the stratosphere – the zone that lies above the tropopause, extending to about 50 km above the Earth's surface.

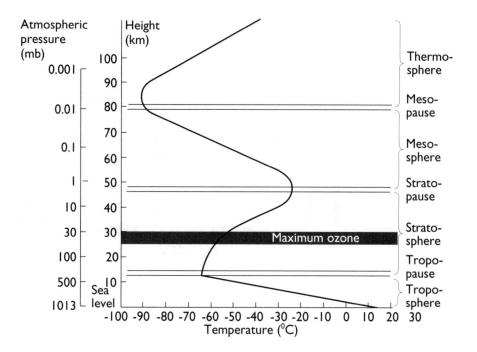

The vertical structure of the atmosphere

atmospheric circulation models are theories put forward to explain the general circulation of the atmosphere. This circulation, in the form of global wind patterns, results from the differential distribution of solar radiation over the Earth's surface. Because year to year changes in global temperature are only very small, the incoming solar radiation must be balanced by outgoing radiation from the Earth. However, this balance varies with latitude. There is a net surplus of radiation at lower latitudes between 35°S and 40°N and a net deficit on the poleward sides of these latitudes. If the tropical regions are not getting progressively warmer, and the higher latitudes are not getting colder, there must be a continuous transfer of energy from the tropics to polar latitudes. Global wind patterns and ocean currents transfer this heat.

The first model of part of the circulation was suggested by Edmund Halley (1686) with warm tropical air rising and spreading toward the poles at high altitude with a return flow toward the equator at low level. George Hadley modified this in 1735 to include the effects of the Earth's rotation, deflecting winds toward the right in the northern hemisphere and to the left in the southern hemisphere to produce the NE and SE Trade Belt. This part of the system is referred to as the *Hadley cell*; it was a single-cell model, which did not extend into the westerly wind belts of the middle latitudes. Further improvements were made by William Ferrel in 1856 when he developed a three-cell model giving a reasonably complete system of global winds. This was further refined by Carl-Gustaf Arvid Rossby in 1941. Although this tricellular model forms the basis of our understanding of global circulation, it does not allow for the influence of depressions/anticyclones or high-level *jet streams* in the redistribution of energy. More recent approaches, known as wave theory models, have been developed to explain the behavior of the upper air westerly air streams (*Rossby waves*) and jet streams.

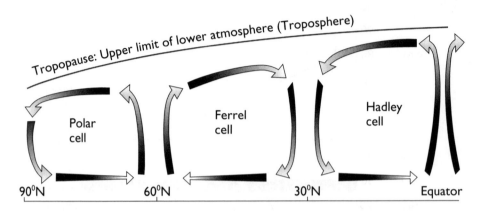

Tricellular model

atmospheric particulates: the name given to constituents of the *atmosphere* that are solid rather than gaseous. They are generated from a number of sources. In rural areas they are made up mainly of soil and sand (also volcanic dust and salt from sea spray). In urban areas they derive mainly from power stations and vehicle exhausts, particularly those burning diesel fuel. In such areas, particulates cause health problems such as asthma and eye irritation and are responsible for the increased occurrence of fogs and *smog*.

atolls: see *coral reef*.

attitudes are sets of beliefs that predispose a person, group or organization to perceive and act toward people, environments and situations in a particular way.

autonomy is the right of self-government. Separatist groups, for example the Basques, Quebecois, Bretons and Scottish Nationalists, seek autonomy from their respective national governments.

avalanche is the term applied to rapid downslope mass movements of rock, ice and snow. They occur most commonly in winter and spring when snow melt helps to lubricate snow and rock debris. Snow avalanches form from recent falls of uncompacted snow on steep slopes, or from partially thawed layers of older snow.

azonal soil is an immature soil that lacks a well-developed profile because it has been little affected by soil forming processes. There are no well-developed soil horizons. These soils are strongly influenced by factors such as *parent material* or relief and in the American 7th. Approximation Classification they are divided into lithosols, regosols, and alluvial *soils*. Lithosols develop at high altitude or in exposed sites where a resistant parent material (scree) reduces the impact of weathering and where steep slopes promote downslope movement, which prevents the formation of horizons. Regosols form on unconsolidated materials (sand dunes and volcanic ash) where profiles are disrupted by the accumulation of fresh material. Alluvial soils (river floodplain or salt marsh) are subjected to periodic flooding, the addition of superficial deposits and poor drainage, which restrict soil processes.

backshore is the name given to that part of a beach that lies above the high water mark and is usually beyond the reach of wave action.

backward integration: a type of vertical integration in which a company gains control of activities upstream from it, i.e. those firms that provide the raw materials to the firm in question. For example, a textile manufacturer could buy up a company that grows cotton.

backwash: the term has two meanings:

- the movement of water back down the beach toward the sea after the *swash* has reached its highest point
- the movement of resources from peripheral areas to a core, particularly to be seen in developing countries. Such resources include raw materials, food, people with skills and ideas, and capital from savings. This was described in the core-periphery theories of John *Friedmann* and the work of Gunnar *Myrdal*. Backwash ultimately leads to *polarization*, in which there is uneven development between the core of a country and the peripheral areas.

bacteria are microscopic single-celled organisms. They are key decomposers of the organic matter of the soil. There are many specialist types able to convert mineral compounds into forms that can be absorbed by plants. Bacteria are also involved in the process by which nitrogen in the air is transformed into *nitrate*, which is an essential nutrient for plant growth.

bahada/bajada is a gently sloping plain found against a cliff in a desert area. It is made up of many *alluvial fans*, occurring at the mouths of *wadis*, which have merged to form this larger feature.

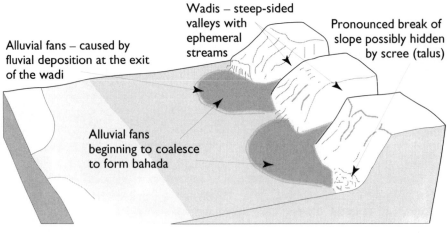

Bahada/bajada

balance of payments: the sum of a nation's income and expenditure on foreign trade. Within the overall balance of payments there is the balance of *invisibles* and the *balance of trade*.

balance of trade is the term used to describe that part of the *balance of payments* account that registers exports and imports of *visible* goods.

bankfull is when the channel of a river contains all the water that it can carry, i.e. its carrying capacity is at a maximum. If a river exceeds its bankfull discharge, then *flooding* occurs.

barchan: a crescent-shaped sand dune that is concave in the direction toward which the wind is blowing, and has a pair of horns that project downwind. They are associated with arid areas that have winds blowing from a constant direction. The windward slope is quite gentle, whereas the leeward slope is steeper (34°). Barchans can be as large as 400 m wide and 30 m high, although many are smaller. The horns tend to move faster than the center, but they move into the shelter of the main body of the dune. Thus a shape is created that can only be altered by a shift in the direction of the wind. Sand is driven up the windward slope and is deposited near the top, thus steepening the leeward slope. When the angle of the leeward slope exceeds 34° slumping takes place and the barchan moves forward. In Peru, barchans have been measured as moving at 25–30 m a year. Although barchans have a distinctive shape, their occurrence is relatively rare.

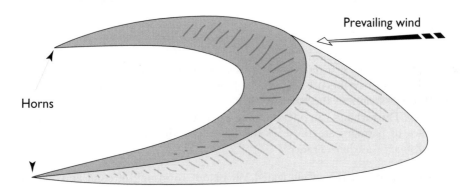

Prevailing wind

Horns

A barchan

bar chart: a diagram consisting of a series of vertical rectangles rising from a horizontal axis. All of the rectangles are of the same width. The height of the bar is proportional to the quantity represented. The horizontal axis is often used to show time intervals (e.g. years, months), and the vertical axis to show values such as amounts and frequencies. All bar charts should begin at zero on the vertical axis.

There are various forms of bar chart:

- a standard bar chart
- a multiple bar chart
- a divided bar chart (often using percentage data)
- a divergent bar chart

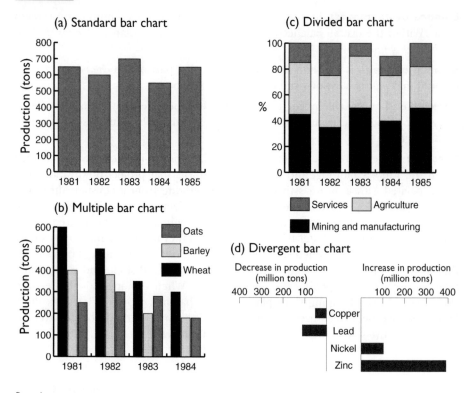

Bar charts

barrage: a dam constructed across an *estuary* with the intention of harnessing *tidal energy*. Barrages have been built across estuaries with a large tidal range. The tides pass through sluices in the barrage, each of which contains a turbine to generate electricity. The turbines can operate using each of the incoming and receding tides. Examples of barrages are those in the Bay of Fundy (eastern Canada) and the Rance Estuary (northern France).

barrier beach: a series of elongated low islands consisting of coral and/or sand lying parallel to a coastline. They therefore have ocean on one side, and marsh and lagoons on the landward side. Sand is deposited by the sea under normal low energy conditions. Wind may then move the sand to build dunes further up the beach, which in turn become colonized by stabilizing plants. Breaks in the islands are maintained by the scour of tidal currents and rising tides spilling into the lagoons behind. Storm waves may wash over the beach, but after a while the beach and lagoon reestablish themselves during low energy conditions.

Some barrier beaches may stretch across a bay connecting one headland with another. Examples of barrier beaches are found at regular intervals along the Atlantic coast, for example at Miami Beach. The formation of the latter is attributed to a postglacial rise in sea level, which flooded the land behind a preexisting stretch of sand dunes.

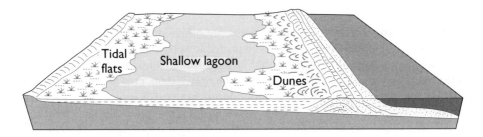

Barrier beach

barrier island: one element of a *barrier beach.*

bars: deposits of sand and shingle situated some distance from a coastline. They usually lie below the level of the sea, only appearing above the level of the water at low tide. There are two theories on where and how they form:

- in shallow seas where the waves break some distance from the shore
- where steep waves break on a beach, creating a strong backwash that carries material back down the beach, forming a ridge.

When a bar begins to appear above the level of the sea for most of the time, it becomes a *barrier beach/island.*

basal sapping occurs when *erosion* is concentrated along the base of a slope causing undercutting and retreat of the slope. Headwall retreat in a *cirque* (corrie) may involve basal sapping; weathered rock fragments become incorporated in the ice and are removed. It can also occur at the foot of a *scarp* where the accumulation of groundwater produces a concentration of moisture, which leads to increased *chemical weathering.*

basal slipping or sliding is an important process of glacier flow, particularly in temperate glaciers. In areas with milder summers, which allow melting to occur, the temperature at the base of the glacier is close to the *pressure melting point* of the ice. As the ice begins to move, there is an increase in pressure and friction with the underlying bedrock. This in turn increases the temperature and causes the basal ice to release meltwater, which acts as a lubricant between the base of the glacier and the bedrock, enabling the glacier to move more rapidly.

base flow is that part of a river's *discharge* produced by *groundwater* seeping slowly into the bed of the river. It is the main contributor to a river's flow during dry conditions. It is relatively constant in amount, although it may increase slightly during a wet spell of weather.

base level is the lowest limit to which *erosion* can take place. For rivers, the ultimate base level is regarded as sea level. Exceptions occur when rivers drain into inland basins that are themselves below sea level. When the course of a river crosses a band of more resistant rock it can produce a local base level. This is a temporary feature; in time the river will erode the resistant layer and the local base level will be removed.

base year: when time series data are put into *index number* form, the year chosen to have a value of 100 in the index series is called the base year.

battery farming is an *intensive* type of *commercial* animal production in which poultry are reared in specially designed units that enable very high levels of output. The birds are kept in cages and are provided with controlled amounts of water and food delivered automatically. Although the production system generates lower priced food for consumers, animal rights groups oppose these methods.

beach: an accumulation of sand and shingle that often occurs in a sheltered position on a coastline, but may also be found in more exposed conditions where there is a plentiful supply of sediment. The upper part of a beach is often composed of coarser materials such as pebbles, and has a steeper slope (10–20°). The lower part of the beach is composed of sand or mud, with a low gradient (2°). (See also *storm beach* and *berm.*)

beach depletion: the reduction in the amount of sand and/or shingle on a beach. It can result from the process of *longshore drift,* or from the loss of natural replenishment methods by obstructions such as a *groin.* Groins are frequently built across a beach to prevent the movement of sediment. The sediment builds up on one side of the groin until it is high enough to overtop it. However, this prevents sediment moving along the coastline and protecting it. Such areas then become liable to erosion by the sea, which can cause significant problems.

It is also thought that other natural replenishment methods are not as active as they once were:

- many of the world's beaches are composed of material deposited on the *continental shelf* at the end of the last glaciation, when sea level was 120 m, lower than today. This material has been transported shorewards by rising sea levels since that time. However, since sea levels have become much more stable, the movement of such material has now ceased
- the extraction of large quantities of this material from offshore zones for commercial purposes has led to a further reduction in its availability
- many rivers do not transport the same quantities of gravels as they once did during the latter stages of the last glaciation
- the colonization of beaches by vegetation has also stabilized much sediment, thereby removing it from the system.

beach nourishment: the replacement of beach material that has been removed by *beach depletion.* There are three main sources of such material:

- land-based sand and gravel pits
- dredged deposits from offshore
- recycled material from other beaches.

In some extreme cases, sand and shingle may be regularly transported from one end of a beach to the other.

Beaufort scale: a numerical system for identifying and measuring the speed of a wind by examining the effects of the wind on natural and structural features. A version of the scale is:

Scale force	Wind name	Speed (km/hr)	Wind effects
0	Calm	0–1	Smoke rises vertically
1	Light air	1–3	Smoke deviates to show direction
2	Light breeze	4–11	Leaves rustle, wind vanes move
3	Gentle breeze	12–19	Leaves move, small flags extend
4	Moderate breeze	20–29	Raises dust, small branches move
5	Fresh breeze	30–39	Small trees sway, waves form
6	Strong breeze	40–50	Large branches move, whistling effect
7	Moderate gale	51–61	Whole trees move, difficult to walk into
8	Fresh gale	62–74	Small branches break off trees
9	Strong gale	75–86	Slight structural damage, e.g. slates blown off roofs
10	Whole gale	87–101	Structural damage likely, trees blown down
11	Storm	102–115	Widespread damage
12	Hurricane	116+	Devastation

bedding plane: the line of division between each layer (strata) in a *sedimentary rock*. It usually indicates where one phase of deposition has ended and another begun. Bedding planes can provide openings or cracks along which *weathering* processes can operate.

bedload: the larger particles (sands, gravels and pebbles) that are forced to roll, slide or bounce along the bed of a river by the force of the moving water in that river.

beneficiation is the concentration of raw materials, usually metallic ores, close to where they are extracted. The waste content in the ore is reduced, as is its volume. Consequently, the costs of transporting the ore to the processing plant are lowered considerably.

Benioff Zone: the boundary between an oceanic plate that is undergoing *subduction* beneath an overriding continental plate. It is a sloping plane (usually between 30° and 60°). The sinking oceanic plate is much colder than the crust into which it is sinking, and sudden stresses along the Benioff Zone may trigger *earthquakes*. It is also within the Benioff Zone that remelting of the subducting oceanic plate may take place. The andesitic lavas created in this way may rise to the surface to create a volcanic *island arc*.

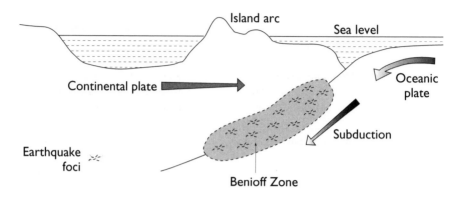

Benioff Zone

Bergeron–Findeisen process is one of the ways in which it is suggested that raindrops can be formed. These two meteorologists noted that in air temperatures of between −5°C and −25°C, supercooled water droplets and ice crystals can coexist. When this happens, the air is oversaturated in terms of the ice crystals and vapor is then *sublimated* onto their surfaces. To compensate for the removal of this vapor, the water droplets evaporate into the air and this increased vapor is again sublimated onto the ice surface. The ice crystals therefore grow at the expense of the water droplets. Eventually, as the ice crystals lock together to form snowflakes, they become large enough to fall. If the flakes melt in the higher temperatures nearer the ground, they form drops of water. This process is the basis of the artificial creation of rain through *cloud seeding*.

bergschrund: a large crevasse near the upper limit of a *cirque* glacier formed as the ice pulls away from the head wall. The bergschrund hypothesis suggests that *freeze-thaw action* at the base of the crevasse helps to break up the rock of the head wall, producing fragments that become incorporated into the glacier. These are transported away, subjecting the head wall to a form of *plucking* that helps to maintain the steepness of the back wall. Recent research on temperature variations within the bergschrund has indicated that the changes may be too small to produce freeze-thaw action and this has led to a reassessment of the ideas; for many geographers the hypothesis is no longer acceptable in its original form.

berm: a ridge of material that is found running across the back of a beach (*backshore*). It marks the highest line on a beach that the waves have reached at a previous high tide.

best contemporary practice is the name given to the policy used by many large *transnational firms* involved in the mining industry, in which all companies within these groups are urged in their operations to "use techniques appropriate to the situation, taking into account relevant health, safety, environmental, social and economic factors." A company that uses such practices is the mining conglomerate, RTZ.

best-fit line: on a *scatter graph,* a line can be drawn as close to all the plotted points as is possible, indicating the *trend* in the pattern that is under investigation. Some points may well lie at some distance from this line and are therefore anomalous, these being known as *residuals.* The line is usually drawn by eye. If all the points fit exactly onto the best-fit line, then the *correlation* between the two variables is perfect.

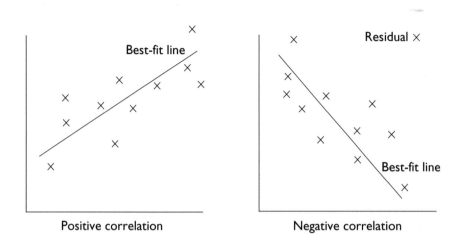

Positive correlation   Negative correlation

Examples of scatter graphs with best-fit lines

bias occurs in *sampling* when there is some distortion or error in the sampled data. The sample is not representative of the total (parent) *statistical population,* i.e. it does not have the same characteristics as the base population. Bias may result from poor choice of sampling method or when an insufficient number of samples have been selected.

bid-rent theory: in an attempt to explain land use within urban areas, this theory states that in a free market the highest bidder will obtain the use of the land. The highest bidder is likely to be the one who can obtain the maximum profit from that site and so can pay the highest rent. In an urban area, competition for land is seen as being greatest at the center because of the accessibility of central areas. In more recent times, though, sites on the edge of urban areas have become more sought after and have seen the growth of industrial areas and out-of-town malls.

bifurcation ratio: the relationship between the number of *streams of one order* of magnitude in a *drainage basin* and those of the next highest order. It is obtained by dividing the number of streams in one order by the number in the next highest order. As the ratio is reduced so the risk of *flooding* within the basin increases.

Worked example: bifurcation ratio

$$\frac{n1 \text{ (number of first order streams)}}{n2 \text{ (number of second order streams)}} = \frac{28}{6} = 4.66$$

Having found all the bifurcation ratios within the basin:

$$\frac{4.66 + 3.00 + 2.00}{3} = 3.22 = \text{bifurcation ratio for the basin.}$$

bilharziasis: a disease caused by a parasitic flatworm, the blood fluke. The worm originally develops in a snail host within fresh water and enters the bloodstream when people work, bathe or swim in the water that contains the snails. The disease is found all over the world, but is particularly prevalent in Africa. The *WHO* has estimated that around 200 million people are infected with the worm, making the disease humanity's most serious parasitic infection after *malaria.*

biofuel is that part of the *biomass* that can be converted into energy. At its simplest, this involves the burning of fuelwood, dung and crop residue for cooking and heating, but this is mostly inefficient. Modern techniques also involve burning the fuel but gasifying the biomass rather than immediately burning it. In developed countries, biofuels also include *landfill* gas and municipal waste. Brazil has developed a system that converts sugar cane to ethanol, which by 1990 had provided that country's drivers with over half of their transportation fuel.

biogas: the process by which methane gas is obtained from animal dung, human excreta and crop residues. The gas can then be used either directly as a cooking or engine fuel or, in a high-technology application, to fuel high-efficiency gas turbines in order to generate electricity.

biogeochemical cycles: the way in which various chemicals (e.g. carbon, nitrogen, phosphorus) are circulated around the *ecosystem* and recycled continually. At its simplest, each cycle consists of plants taking up chemical *nutrients*, which they pass on to both grazing animals and carnivores. As animals at each *trophic level* die, they decompose and the nutrients are returned to the soil. These cycles can operate on land, in the sea, or in the air.

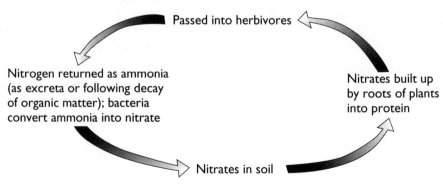

An example of a biogeochemical cycle – nitrogen

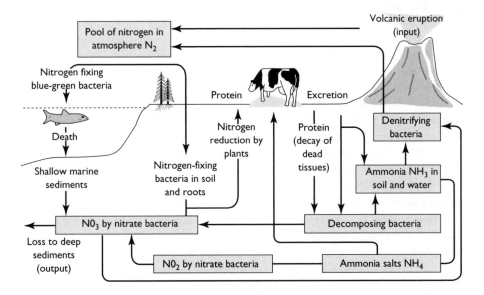

Detail of the nitrogen cycle

biological controls are natural predators, parasites, bacteria and viruses used to control weeds and pests in agriculture. This approach reduces the need for chemical controls that may accumulate in the *food chain*. However, effective control is rarely achieved solely through biological methods and some application of pesticides/herbicides is usually needed.

biological oxygen demand (BOD): the amount of dissolved oxygen needed to enable the decomposition of organic material in polluted water. It is measured in milligrams of oxygen per liter of water at 25°C. Polluted streams have low values of dissolved oxygen content and thus have a high BOD. Heavily polluted water courses may have a BOD of 350 to 400.

biomass: the total amount of organic material both above and below the ground surface and in water bodies. Plants and vegetable matter, both living and decaying, comprise the greatest bulk of the Earth's total biomass. Animal biomass is small in comparison and mostly consists of microorganisms that live in the soil.

biome: a naturally occurring organic community of plants and animals. Each biome derives its name from the dominant type of vegetation found within its physical environment or habitat (e.g. savanna, grassland, coniferous forest) and consists of the *ecosystem* of plants, soils and animals.

biosphere reserve: a large area set aside by a government in order to conserve the area's special ecological characteristics and to protect certain species.

biotechnology is the application of biological knowledge and research to technological development. It has emerged as a high-tech industry, which uses the properties of living cells in specialized areas such as the discovery and production of vaccines and antibiotics.

biotic factors are those that result from the action of living organisms (plants and animals) influencing plant growth and distribution. Plants can modify the physical conditions of a habitat by providing shade and shelter; they alter light intensity, temperature and humidity conditions and they can reduce wind speeds near to ground level. In this way they can create their own particular microclimate, and through their use of water, mineral nutrients and the return of organic matter they also influence the condition of the soil. The development of a given species in a site depends upon its ability to compete for space, light, water and soil nutrients. The term also includes factors related to the activity of animals within the habitat. Many plants depend on animals for seed dispersal and crosspollination of their flowers. However, plants are also basic food producers and may be damaged by animals that feed on them removing leaves or bark.

bipolar test: a method of determining *attitudes* toward a particular phenomenon. It is based upon two extremes of attitude that are said to be "poles apart" or "bipolar" views. It is possible to ask people to judge a phenomenon by scoring somewhere between these two poles on a grid like the one below:

Bipolar scoring chart for...industrial site

Location..Time...

Date......................................

		5	4	3	2	1	
The layout is	ATTRACTIVE	X					UGLY
The amount of foliage is	POOR				X		EXCELLENT
The traffic flow is	CONGESTED			X			LIGHT
This place is	QUIET				X		NOISY
Smells are	PLEASANT				X		OFFENSIVE
Street and premises are (e.g. litter, graffiti)	DIRTY				X		CLEAN

NB All the negative words should not be placed on one side of the chart as this may bias responses.

Bipolar scoring chart

birth control programs are designed to limit births and therefore lower the *birth rate*. These programs are typical of *Third World* countries, particularly in Asia, such as China, Sri Lanka and Singapore. The program in Singapore was typical of many in having several strategies:

- to establish family planning clinics and to provide contraceptives at minimal cost
- to advertize in the media the need for, and the advantages of, smaller families

- to legislate to allow abortions and sterilization
- to introduce social and economic incentives such as paid maternity leave, income tax benefits, housing priority, cheaper healthcare and free education, which would cease if family size grew.

birth rate is a measure of a country's fertility. It is expressed in the number of live births per 1000 people in one year.

blockfields are extensive sheets of large angular rock fragments formed by frost action in *periglacial* areas. When areas of exposed bedrock are subjected to fluctuations of temperature through freezing point, *freeze-thaw* processes are very effective in producing accumulations of shattered debris. These are also called felsenmeer. Where blocks have moved downslope to form lines of angular debris they are called block streams.

blocking high: an anticyclone that prevents depressions from following their normal paths by diverting them around the edge of the high pressure. They occur when an *anticyclone* breaks away from the Azores High and establishes itself in more northerly areas (latitude 50°N–70°N) for several days. Once in place, the system may remain for several weeks in extreme cases, altering the usual weather pattern. The effects depend on the season, the position of the blocking high and the length of time it persists.

blue collar: a term used to describe employment in manual work.

bluff: a steep slope resulting from lateral erosion by a river cutting back interlocking spurs. This may create a bluff line on the edge of a wide flood plain.

BOD: see *biological oxygen demand.*

bog: blanket bogs develop under conditions of high rainfall in areas that are too wet for tree development. They are areas of waterlogged, spongy ground dominated by herbaceous plants such as bog moss (sphagnum) and cotton grass, bell heather, bog myrtle and rushes. These decay slowly to form highly acidic peat with *pH values* just above 4. The acidity is due to the high rainfall maintaining a downward flow in the ground water. This creates *anaerobic* conditions in which bacteria cannot survive and organic decomposition is extremely slow.

Raised bogs are so called because their central regions are at a higher level than their margins. They are commonly found on silted glacial lakes or on estuarine muds that were originally colonized by *haloseres*. This is due to the fact that after the bog plants had invaded they continued to develop, producing more and more peat, which accumulated more rapidly in the central area.

boom irrigation: water is introduced to the field using a sprinkler that rotates around a central pivot. This creates a pattern of circular higher yielding crop areas separated by lower yield or fallow areas.

boreal: this term means northern. It is an alternative name for the largely coniferous forests that extend across large areas of North America and Europe between latitudes 45°N and 75°N. These boreal forests experience severe climatic conditions, with cold winters and short summers. The growing season is less than six months, and may be as low as three months on the northern limit. The trees, which are mainly evergreen species such as pine, fir and spruce, have adapted to the conditions with flexible branches to shed snow and needle-like leaves to reduce water loss through transpiration. This is particularly important in the winter when freezing of the soil restricts availability of water; the trees experience a physiological drought.

The term is also used for the Boreal climatic period, which extended from approximately 7000 years *BP* to 5500 years BP. This was a period that followed the cold and wet Pre-Boreal when pine and birch trees dominated the vegetation. The Boreal was cold but drier. As a result the species favoring "wet" conditions, such as birch and pine, declined and trees like elm and hazel reached their peak. Oak trees appeared during this stage and became increasingly important as the climate continued to get warmer. They reached their dominance in the warmer Atlantic period that followed the Boreal.

Boserup, Ester: a Danish economist who in 1965 suggested a theory on the relationship between population and resources that opposed the views put forward by Thomas *Malthus*. Whereas Malthus argued that food supply acts as a ceiling to population growth, Boserup suggested that in a preindustrial society an increase in population stimulates a change in agricultural methods that results in greater food production. If groups do not produce more food then starvation follows, an example of the saying "necessity is the mother of invention." From her studies on different agricultural systems, Boserup suggested that farming became more intensive as population pressure increased; fallow periods became shorter, with one crop harvested a year and a fallow period of only a few months. With further population growth multi-cropping is stimulated as the most intensive system.

Boswash is the name given to the urban area that extends from Boston through New York to Washington along the Atlantic coast. It is sometimes referred to as Bosnywash.

The term "megalopolis" was originally used to describe this area with one metropolitan conurbation coalescing with another, but the term is now used for any continuous built-up area of more than 10 million inhabitants.

bottom-set beds consist of the finest clays deposited in a *delta*. They are deposited in near horizontal layers at some distance from the mouth of a river. The settling rates of such clays is very slow, but in salt water the particles are subject to *flocculation* causing them to sink to the sea bed.

boulder clay is the unsorted and unstratified debris stranded or deposited over a landscape by the action of ice. It is composed of fragments of rock of all shapes and sizes, ranging from large boulders to fine clay particles. The stones in boulder clay tend to be angular or subangular. The composition of boulder clay in an area reflects the character of the rock over which the ice has moved to that area. It is also known as *till*.

BP: an abbreviation for "before present." It is used for stating the age of relatively recent deposits from the glacial or postglacial periods where dates given as BC would not present such an accurate impression as to how long the deposit had been formed. It is applied particularly to the results of *carbon dating* where absolute dates plus or minus margins of error are given.

braiding is when a river is forced to divide into several channels with islands separating them. It is a feature of rivers that are supplied with large loads of sand and gravels. The banks formed from these materials are unstable, and consequently the channel becomes very wide in relation to its depth. The river becomes choked, with several sandbars and channels constantly changing their location. Braiding occurs in a number of environments where there are rapidly fluctuating discharges:

- semiarid areas of low relief that receive rivers from mountainous areas
- glacial *outwash* plains
- *periglacial* areas underlaid with *permafrost.*

Brandt Report: published in 1980 to highlight the growing gap in social and economic development between the developed countries of the world, *The North,* and the less developed countries, *The South.* It was compiled by an independent group of statesmen headed by Herr Willy Brandt, the former West German Chancellor. The Report discussed a range of issues: disarmament, political corruption, violation of human rights, overpopulation, world health, industrialization, world trade, environmental pollution, urbanization, and forms of communication between countries. It concluded that "The North" and "The South" were mutually dependent upon each other, and it warned against "The North" establishing economic barriers against the growing industrialization of "The South."

break of bulk: a point where cargo is unloaded from some form of bulk carrier and is transferred to smaller units of transportation for further movement. It applies where the mode of transportation changes, e.g. sea to land. These are attractive points in terms of economic location because they offer potential savings in transportation costs. If raw materials are processed at the break of bulk point there is no need to transfer materials from sea-going to land-based transportation – they can be unloaded directly into the processing plant without further shipment costs. In this way the expensive unloading and reloading costs can be reduced. Heavy industries that process imported raw materials find tidewater break of bulk locations advantageous.

Bronze Age settlement dates from 1900 BC to 500 BC. Common forms of evidence of human settlement at this time are hill forts, burial mounds, barrows, remnants of hut circles and ancient field patterns. The hill forts are the most obvious, consisting of grassy ramparts and ditches encircling hilltops and coastal promontories. Entrances within the ramparts and ditches can be identified in most cases.

brown earth: a type of soil associated with the northwest European type climate and *deciduous woodland.* The considerable amount of leaf litter that accumulates in the autumn is decomposed relatively quickly by a range of soil organisms to create less acidic *mull* humus. This is incorporated into the upper horizons of the soil by earthworms, giving it a dark brown color. There is a downward movement of soil water due to *precipitation* exceeding *evapotranspiration,* but the degree of *leaching* is limited. However, some clay particles are washed down through the *soil profile* (*lessivage*), and the lower horizons become enriched with clay. The different horizons in the soil merge gradually due to the active mixing by soil fauna. The soil tends to become lighter in color downward through the profile. Brown earth soils are potentially fertile, though they benefit from liming.

Brown earth soils on *parent material* such as granite and sandstone tend to be more acidic, whereas those on calcareous rocks (limestone, chalk) are less acidic.

Burgess, Ernest Watson: an American sociologist who in 1924 proposed the *concentric urban model* based on his work in Chicago. Burgess employed ecological factors to explain the spatial variations within a city. In ecology the emphasis is on the interrelationships between living things and their environments. He was concerned with the factors of invasion, competition, dominance and *succession* as used in biology to explain the distribution of plants. In short, Burgess substituted people for plants.

burglary is one of the hazards of the human environment, particularly the urban environment. The nature of the hazard concerns not only the removal of property but also the invasion of privacy and the damage that may be caused to the dwelling.

The frequency of occurrence is not easy to estimate because:

- if police reports are the source of information, not all burglaries are reported
- if the source is insurance company data, not everyone has insurance
- local newspaper evidence is also flawed.

The distribution of the hazard affects strategies for dealing with burglary, such as policing policies, *neighborhood watch* schemes and modifications to the building structures in crime-ridden areas. Areas with protection schemes may see a decline in burglary, but the crime may have been displaced to another less well-protected area.

business cycle: the regular pattern of upturns and downturns in demand and output within the economy that tend to repeat themselves every five years or so. The causes of this cyclical pattern to economic activity are not fully known, but are partly explained by:

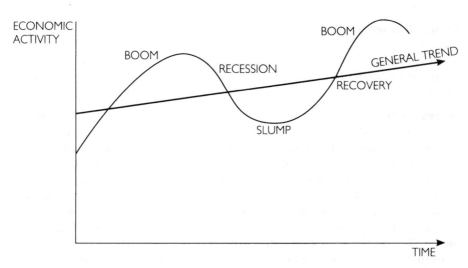

The phases of the business cycle

- the bunching of investment spending, which, by definition, does not need to be repeated for several years
- government policies that aim for rapid growth just prior to elections, leading to inflation and therefore the need to constrain the economy after elections
- confidence in the future, which means that firms will invest and expand, contributing to the economic upswing. Alternatively, if firms foresee an economic slowdown in the economy they cease to expand and postpone investment plans.

business park: a term adopted by property developers in order to attract firms needing office accommodation rather than industrial units, as well as *hi-tech industries.* Some business parks may include leisure activities such as golf courses and horseback riding centers.

Buys Ballot, Christoph Hendrik Diederik: a Dutch scientist who, in 1857, proposed a law stating that if an observer stands with his back to the wind in the northern hemisphere low pressure lies to the left. In the southern hemisphere the reverse holds true. This can be used to predict wind direction in a weather system if the position of high and low pressure is known. The law is based upon the effects of the *Coriolis Force.*

C

Cairo Population Conference 1994: a conference convened by the *United Nations Fund for Population Activities*, which was to take stock of the current population situation and attempt to reach a consensus about population policy for the future. Heated debate was raised about family planning, especially between those promoting such programs and the Vatican and certain Muslim nations. Many Third World countries showed a determination to reduce their birth rates and others stressed the importance of the empowerment of women in order to gain greater access to family planning. At the end, a consensus was reached, even if it was not possible for the nations to agree on population targets either for the world or for individual countries.

calcareous soils are those that are derived from a *parent material* such as *limestone* and *chalk*. They are naturally alkaline.

calcification occurs in soils in areas of low precipitation, where rates of evaporation are high, and where there is a water deficit for a large proportion of the year. When rain falls it is sufficient to penetrate the upper horizons, dissolve some calcium and percolate downward. However, there is insufficient rainfall to perform *leaching* of the soil effectively, and soon the water is evaporated, leading to the deposition of calcium carbonate.

calorie intake: a kilocalorie is a measurement of the energy-producing capacity of food and is a way of showing the *diet* of populations in different parts of the world. It has been calculated that the average adult in temperate latitudes requires 2600 kilocalories per day compared to 2300 for people in tropical areas. The average daily consumption is 3300 in the *First World*, but only 2200 in the *Third World*. Modern methods of measurement have replaced calories with megajoules. (See *diet*.)

calving is the process of *ablation* by which small masses of floating ice break away from an ice sheet or glacier. This can produce icebergs if the edge of the ice cap extends into the sea or smaller masses if a glacier terminates in a lake.

canopy: the highest layer of foliage in a woodland formed by the crowns of fully developed trees. A well-developed canopy can significantly reduce light intensity, which will restrict the growth of smaller trees and shrubs.

CAP: see *Common Agricultural Policy*.

capacity: the largest amount of *load* that a river can carry for a given velocity. Research has shown that a river's capacity increases according to the third power of the velocity of that river, i.e. if a river's velocity doubles, then its capacity increases by eight times (2^3).

capillary action is the movement of water upward through a substance. It is caused by the adhesive attraction that water molecules have for the walls of the surrounding surfaces. The smaller this space, the greater the degree of capillary action.

capital represents the finance invested in a company either to start up that business or for production and expansion. Capital is obtained either from shareholders (share capital) or from lenders (loan capital). Capital can also said to be fixed. This is the investment in buildings and equipment and is not mobile compared with money capital. Some geographers argue that there is a third form of capital, social capital. This is represented by housing, hospitals, schools, shops and recreational amenities, which may attract a firm, particularly its management, to an area.

capitalism is the social and economic system that relies on the market mechanism to distribute *factors of production* in the most efficient way. Most of the capital or wealth is owned and controlled by individual people or companies rather than by the state or government.

carbon dating (radio-carbon dating) is a means of determining the age of pre-historic organic remains (e.g. wood, bone) up to about 50,000 years *BP*. It is based on the fact that radioactive carbon or carbon-14 decreases at a known constant rate after the death of the organism (half-life of $5,730 \pm 40$ years – i.e. half of the carbon-14 present will decay during that period).

carbon dioxide: one of the gases that occurs naturally in the Earth's *atmosphere* and usually takes up only 0.03% by volume. Carbon dioxide absorbs long-wave radiation from the Earth and is one of the factors that in the past has kept the temperature steady. In the 20th century, however, carbon dioxide has substantially increased (on some estimates by at least 15% in the last hundred years and it could double by the middle of the 21st century) mainly through the burning of *fossil fuels*, although *deforestation* has played some part. The overall effect of this increase is that more long-wave radiation will be trapped and therefore lead to an increase in the atmospheric temperature, known as the *greenhouse effect*.

carbon tax is a method of raising energy prices by increasing the revenue obtained by a government from gas and heating oil sales. Such taxes are one of the ways in which it is hoped that the use of *fossil fuels* will be reduced. Several developed countries are contemplating such legislation, Switzerland having already implemented such a measure.

carboniferous limestone: the calcareous rock laid down in the geological period of that name (280–345 million years *BP*) that gives rise to its own particular type of scenery known as *karst*. Major features associated with this limestone include underground drainage systems, limestone pavements, swallow holes, and *dry valleys*. The rock has developed its own particular type of scenery because:

- it is found in well-defined, often thick, beds that are almost horizontal and are well jointed at right angles to the *bedding planes*
- calcium carbonate is soluble through carbonic acid in rainwater, which combines with other acids from upland vegetation
- it is pervious but not *porous*, which means that water can pass along the bedding planes and down the *joints*, but not through the rock itself.

Carboniferous limestone in the USA is found primarily in the Mississippian rock system, which extends from West Virginia and along the Mississippi River in Missouri, Illinois, and southeastern Iowa. (See also *limestone*.)

carrying capacity is the largest population that the resources of a given environment can support. The term has its origins in ecology where the population related to plants, but it is also used to describe the maximum number of livestock that can be supported per unit area. Thomas *Malthus* put forward the concept of a population ceiling where saturation level is reached when the population equals the carrying capacity of that environment.

cartel: the name given to a group of producers who make an agreement to limit output in order to keep prices high. In order to do this they must control a large proportion of the output, and they must agree on levels of production. Probably the best-known cartel is the *Organization of Petroleum Exporting Countries (OPEC)*. The major problems with cartels are:

- if they do force a high price it will encourage other producers to enter the market
- the members of the cartel may cheat by secretly producing more than laid down in the cartel agreement in order to gain more revenue.

In most countries cartels are illegal because of their potential to exploit customers.

cartography is the science and skill of map and chart production.

cash cropping is the growing of crops for sale as distinct from a crop grown for consumption by the farmer and his family. Cash cropping operates well when there are large domestic markets, opportunities for foreign trade and well-developed transportation systems.

Cassa per il Mezzogiorno: the "Fund for the South," a body established in 1950 to assist the economic development of the south of Italy – the Mezzogiorno. The aim was to provide the south with necessary basic structures such as roads, drainage and services, to assist economic development. Aid was given to agriculture through the breaking up of large estates (*latifundia*) into smaller holdings, the

The Mezzogiorno

construction of irrigation schemes, and various incentives. The Cassa concentrated later on the concept of *growth poles* throughout the region. Despite the injection of large amounts of capital the Cassa has only been partially successful in raising the economic profile of the south and the region remains an area of persistent out-migration.

catastrophism is the belief that the Earth's features are produced by sudden catastrophic events rather than by slow, more continuous processes such as plate movement, weathering, erosion and deposition. Until the middle of the 19th century, the recognized explanation for geological deposits was related to the deluge theory linking these materials to the biblical record of Noah's flood. Although the term is outmoded, large events, which are very rare, may produce greater physical changes over a long period of time than the common day-to-day processes.

catena is a sequence of soils down a slope where each soil type is different but linked to its adjacent types. Catenas illustrate the way in which soils can change on a slope where there are no real changes in climate or the *parent material*. Catenas develop over a long period of time and are therefore best established in areas with stable environments. Parts of Africa, particularly East Africa, show good catena development.

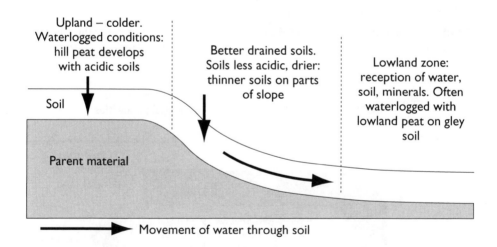

Typical soil catena in upland Britain

cation exchange capacity is the ability of a soil to retain cations (positively charged ions), particularly those of calcium, magnesium, potassium and sodium. Sandy soils have a much lower ability to do this than clay soils, which in turn have a lower capacity than does *humus*. This effects the *fertility* of the soil, sandy soils being much less fertile than ones that contain a great deal of humus.

cavitation is a process that takes place in streams flowing at high velocity or under great pressure, for example in *subglacial* channels. Air bubbles form and collapse within the water causing shock waves against the channel bed and sides. This produces a sand blasting effect where particles in the stream are thrown against the margins of the channel, polishing and smoothing the sides and creating pot holes in the bed.

CBD: see *Central Business District*.

CD Rom stands for "compact disk read only memory." It is a method of information storage and recall via a computer. Like a music CD, it can be played and information accessed, but no new material can be added to it. Data such as newspapers and encyclopedia are increasingly being put on CD Rom.

census: a periodic count of the population of a country, administered by the government. In the US a census is taken every 10 years. The next census will take place on April 1, 2000. The aims of censuses have changed considerably over time, with many more questions asked, so that today it constitutes the collection of information for a broad data bank. This information is of use to both the *public* and *private sectors*:

- government and local government use the census data as the basis on which to allocate resources
- nongovernment users may include retailers, advertisers, financial services, property developers, *utilities*.

There has been some research carried out to try to find other ways of collecting data. In Denmark, Holland and Sweden, for example, rolling registers of population are compiled, but there has to be a legal requirement that everyone be registered and that they record all change, such as changing residence, etc.

Central Business District (CBD) of an urban area contains the principal commercial streets and major public buildings and is the center for business and

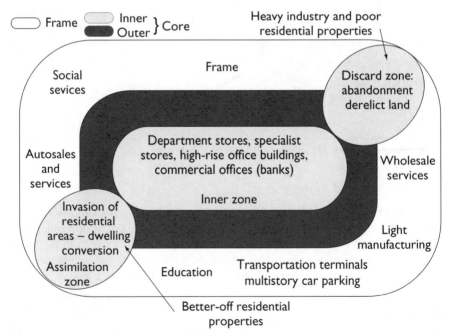

Central Business District

commercial activities. The CBD role was based on its accessibility from all parts of the urban area and as a result contains the highest land values in the area. The CBD is not, however, static. It can be moving outward in some directions (zones of assimilation) and retreating in other parts (zones of discard). In many CBDs *retailing* is declining due to competition from *out-of-town developments* giving a greater emphasis to offices and other services. In a sizeable urban center there is often segregation of different types of businesses within the CBD to form distinct quarters. Retailing tends to separate from commercial and professional offices to form a distinct inner core. The outer core is made up of offices and entertainment centers with some smaller stores. Beyond this, the outer part of the CBD is known as the frame, and contains, among other features, service industries, wholesaling and parking facilities.

centrally planned economies: see *command economies.*

central place theory was an attempt to explain a settlement pattern of *market centers* in a regular order. Perhaps the best-known theory of this kind was developed by Walter *Christaller,* who suggested that there was a pattern in the distribution and location of settlements of different sizes and also in the ways in which they provide services to the inhabitants living within their *spheres of influence.* Each settlement, no matter what its size or level of service provision, was to be regarded as a central place, i.e. a center providing goods and services to its surrounding area.

central tendency: the tendency of values within a set of data to group around a particular value or values. Measures such as the *range, interquartile range,* and *standard deviation* can be calculated to describe the extent to which data is clustered around the *mean.*

CFCs: see *chlorofluorocarbons.*

chalk is a relatively pure form of *limestone* containing a high proportion of calcium carbonate. It is a *sedimentary rock* that forms as a result of the compression of the shells of marine organisms that have accumulated on the floor of a shallow sea. Chalk is a porous rock allowing *groundwater* to pass through pore spaces, although *permeability* is greatly assisted by the presence of joints and bedding planes. Although chalk is a relatively soft rock, when folded or tilted and flanked by other rocks, it can form upland areas because its permeability allows water to pass through, reducing the amount of surface erosion. Where the dip of the chalk is gentle, *escarpments* are formed. Chalk does contain some impurities, notably flints, which are produced from the chemical precipitation of silica. These dry, upland areas have many remnants of occupation by early man, for hard flints were used for tools and weapons.

channel efficiency is measured by the *hydraulic radius,* which is defined as the ratio of the cross-sectional area divided by the length of the wetted perimeter. The higher the ratio the more efficient the channel and the smaller the loss of energy due to external friction. Under normal conditions about 95% of a stream's energy is used in overcoming friction between the water and the margins of the channel and as a result of internal shearing between turbulent currents. Other measures that reflect the efficiency of the stream to move sediment are *capacity* and *competence.*

channel flow is the run-off of surface water within a well-defined channel, as in a river or stream channel. Flow is the *discharge* (Q) measured in terms of the volume of water (in cubic meters) passing a point in the river channel in a unit of time (one second) and is usually given in "cumecs."

channel morphology is concerned with the shape and dimension of channel cross-sections, which can be influenced by a number of factors: the *discharge*, the quantity and caliber of *sediment load*; and the materials that form the bed and banks. Channel shape may be described by the *hydraulic radius*. Where both banks are being eroded channels tend to be broadly rectangular or trapezoidal in shape. Where one bank undergoes more rapid erosion, as in a *meander*, the channel is likely to be more asymmetrical. The shape of the channel can be expressed in the form ratio: $FR = d/w$ (where d is depth of channel and w is width). Generally form ratio decreases downstream as width tends to increase more rapidly than depth. River channels also show great variation in plan; some are relatively straight, some are sinuous, and some are fully meandering. The *sinuosity* can be measured to assess the degree to which a river is meandering. Many channels also have the flow subdivided by depositional material, which produces *braided* patterns.

channel processes operate within rivers to transport, erode and deposit sediment. Energy remaining after the river has overcome friction can be used to transport sediment. This load is transported by three main processes: suspension, *solution* and as *bedload*. Bedload may be moved by *saltation*, when debris is temporarily lifted up by the current and bounced along the bed, and traction, when larger pebbles or boulders are rolled along the bed. The critical values for the pick up and settling velocities of different sized particles are displayed in a *Hjulstrom curve*. Velocity influences the *competence*, which is the maximum size of material the river is capable of transporting, and the *capacity*, which is the total load transported. Erosion of new material, or the wearing down of existing sediment, can be achieved through a variety of processes: *corrasion* or *abrasion*, where the river uses the load to wear away the bed and banks; evorsion, where the force of the water prises away new sediment; *cavitation*, a pressure effect under high velocity flows; *solution* or corrosion, which is particularly effective when the river flows through areas of *limestone* or *chalk*, which are prone to solution by acidic water. The existing load may be worn down by attrition; boulders collide with other material and angular debris becomes progressively more rounded.

chaparral is a type of scrub vegetation found in the area of California with a climate similar to the Mediterranean region. It is a *biome* characterized by short, woody and dense bushes that have sclerophyllous leaves – small, thick, leathery with thick cuticles. Some species have tomentose leaves, that is, they have a covering of fine hairs that restrict wind *evaporation* and help to reduce the rate of water loss by *transpiration*. The main species are evergreen oaks and a drought-resistant conifer, the pinon. Other *xerophytic* bushes and herbaceous plants are to be found, and bare soil occurs between the plants. The vegetation is closely related to the *maquis* of Southern Europe.

chelation is an important biochemical process in soil formation. Mineral ions, particularly those of aluminum, magnesium, iron and calcium can be held in the molecular structure of organic compounds. If the soil has a thick layer of *humus* on its surface, percolating water will contain humic acids and organic compounds that will take up the mineral ions from the soil solids. Elements can be moved from the upper layers and washed down the *soil profile*; this process is termed cheluviation.

chemical waste includes substances that are either toxic, ignitable, corrosive, or irritant and are potentially dangerous to humans and animals. The leakage of such wastes has been linked to a variety of birth defects, cancers, brain damage and blood and nervous disorders. A number of methods have been used to dispose of chemical waste:

- sealing waste in drums and then storing them, but inappropriate labeling is a problem
- using *landfill sites*, but the chemicals may penetrate groundwater supplies
- incineration both on land and at sea, but this releases toxic gases into the atmosphere
- discharging into the sea or rivers, but this does not remove the problem.

(See also *toxic waste.*)

chemical weathering involves the decay or decomposition of rock in situ. It usually takes place in the presence of water, which acts as a dilute acid. The end products of chemical weathering are either soluble and are therefore removed in solution, or they have a different volume, usually bigger, than the mineral they replace. The rate of chemical weathering tends to increase with rising temperature and humidity levels, except in the action of carbonic acid (carbonation) where lower temperatures produce greater rates of weathering on limestone.

Chemical weathering can also occur from the action of dilute acids resulting from both atmospheric pollution (sulfuric acid), and the decay of plants and animals (organic acids).

There are a number of different types of chemical weathering, including *hydrolysis*, carbonation, and *oxidation.*

chernozem: a type of soil associated with the continental interior-type climate and *temperate grassland.* The thick grass cover provides plentiful *mull humus*, which forms a black upper horizon with a crumb structure. There is an abundance of soil fauna, which rapidly decay and incorporate the organic matter into the upper horizons of the soil during the warm summers. However, as the winters are much colder, the process of decay is greatly reduced. The snowmelt in late spring and the early summer rainstorms cause some *leaching* to take place, but in the later hot summer there is a *capillary* upward movement of water. This alternating pattern of soil water movement causes nodules of calcium carbonate to be deposited at depths of about 1 m. The lower horizons are paler due to the reduction of humus content. Chernozems are naturally very fertile.

chi-squared test: this technique is used to analyze the distributions of dots or points. Normally it compares an actual distribution of points with a *random* distribution of the same number of points. First a *null hypothesis* (H_0) is formulated to the effect that there is no significant difference between the observed pattern in the distribution and the expected pattern of distribution (which is usually regarded as being random). The alternative hypothesis to this will be that there is a difference between the observed and the expected patterns, and therefore there must be some factor responsible for this difference.

The method of calculating the value of chi-square is shown in the chart that follows. The letters A to D in the table refer to the areas A to D in the map alongside the table. In the column headed O are listed the numbers of points in each of the areas A to D on the map (the observed frequencies), the total number of points in this case being 20. Column E contains the list of expected frequencies in each of the areas A to D assuming that the points are randomly spaced. In the columns O–E each of the expected frequencies is subtracted from the observed frequencies, and in the last column the result is squared. The relevant values are then inserted into the expression for chi-square, and the resultant value is 2.0.

The aim of a chi-squared test, therefore, is to find out whether the observed pattern agrees with or differs from the theoretical (expected) pattern. This can be measured by comparing the calculated result of the test with its level of significance. There are two levels of significance – 95% and 99%. At 95% there is a 1 in 20 probability that the pattern being considered occurred by chance, and at 99% there is only a 1 in 100 probability that the pattern is a chance one. The levels of significance can be found in a book of statistical tables. If the calculated value is the same or greater than the values given in the table, then the null hypothesis can be rejected and the alternative hypothesis accepted. In the case of our example, however, the value of chi-square is very low (2.0) showing that there is little difference between the observed and the expected pattern. The null hypothesis cannot therefore be rejected. (See also *significance testing.*)

		O	E	O − E	$(O - E)^2$
A	B				
A		4	5	−1	1
B		7	5	2	4
C	D				
C		3	5	−2	4
D		6	5	1	1
		20	20	0	10

$$\chi^2 = \frac{\Sigma(O - E)^2}{E} = \frac{10}{5} = 2.0$$

Chi-squared test

chlorofluorocarbons (CFCs) are chemicals used in foams, refrigerators, aerosols and air-conditioning units. They are held responsible for the destruction of the *ozone layer* between 15 and 25 miles above the Earth's surface. Following growing concern over *ozone depletion*, a United Nations conference in 1987 (the Montreal Protocol) agreed that there should be a 50% reduction in the production of CFCs by 1999. There is still considerable concern that this reduction will not be adequate because of the slow speed at which they disintegrate within the atmosphere. The European Union countries have now agreed to a 100% ban on CFC usage by the year 2000. CFCs are also thought to be responsible for part of the so-called *greenhouse effect*.

choropleth maps are shaded or colored to display varying spatial distributions within administrative areas. They usually show groupings or classes of data, with a shading system or color allocated to each group. Shading should vary from dark to light for high to low values. Choropleth maps should avoid the use of black or white. Black implies completeness and is difficult to write or print over. White implies emptiness, and is often used to represent areas where there is no data.

The main weaknesses with choropleth mapping are that it is very dependent on administrative boundaries, and shows only average values within that administrative area. It is also unlikely that abrupt changes will take place between such areas, as implied by changes in the types of shading at the administrative boundary – they are much more likely to be gradual.

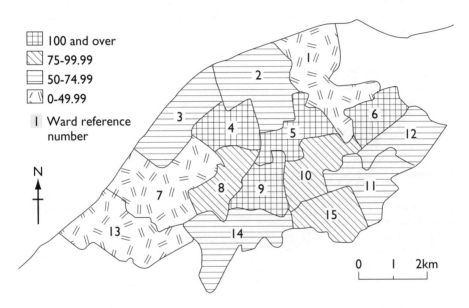

An example of a choropleth map showing the population density of wards of a town

Christaller, Walter published a book in which he demonstrated a sense of order in the spacings of, and the provision of services by, settlements. He suggested that there is a pattern in the distribution of settlements (central places) of differing sizes, as well as a pattern in the services that they provide. His ideas have been called the *central place theory*, and his work has contributed a great deal to the search for order in the study of settlements.

cirque (also known as corrie, cwm): a semicircular hollow high up at the head of a *glacial* valley on the flanks of a glacial mountain. It was formerly a massive collecting ground for ice, which flowed out of it into a glacial valley below. A cirque has a steep headwall to the rear, a bowl-shaped rock basin in its center (sometimes occupied by a small circular lake called a tarn), and a rock lip at its lower end. This lower end of the cirque often marks the point where a stream plunges steeply down into the main glacial valley below.

Cirques develop from an initial hollow in which a snow bank has accumulated. The presence of the snow bank would increase the amount of diurnal and seasonal frost weathering caused by meltwater in the hollow, so that it would gradually be enlarged until it was big enough to hold a small glacier. This glacier would then start to cause *plucking* and quarrying at its head, probably as a result of the downward percolation of water and the *frost-shattering* of the rocks in the headwall. The *rotational movement* of the ice in the hollow also helped to create the rock basin and rock lip.

In North America and Europe most cirques are oriented between northwest and southeast. This is because:

- north-facing slopes receive less insolation and this helps to preserve small glaciers
- the main snow-bearing winds are from the west, and eddies would ensure that snow banks were preserved on the lee slopes, i.e. those facing east.

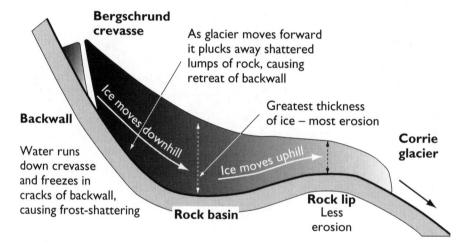

Formation of a corrie/cirque

clapotis occur where the sequence of advancing sea waves coincides exactly with the sequence of reflected waves from a cliff or sea wall. The result is the appearance of a series of "standing" crests and troughs in the waves some distance from the cliff face or sea wall.

Clarke-Fisher model: as an economy develops over time, this model shows in theory how the relative importance of the sectors of employment changes. In pre-industrial times an economy will be dominated by the *primary sector,* but over time people will move from this sector as *manufacturing* develops. To support the growing industrial base and the demands of a more affluent population, there is a need for a whole range of services including transportation, utilities, and consumer and financial services, leading to an expansion of *tertiary* employment. In more advanced economies a *quaternary sector* develops, which encompasses services such as *research and development* and information processing.

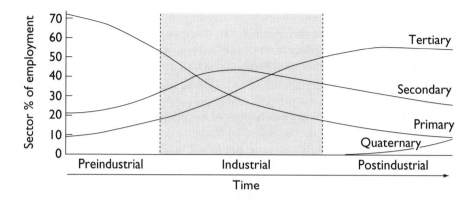

Clarke-Fisher model

class intervals are the dividing lines chosen in order to group data into categories for the purpose of analysis. (See *frequency distribution.*)

climate: the mean atmospheric conditions of an area measured over a substantial period of time. Different parts of the world have recognizable climatic characteristics with distinctive seasonal patterns.

climatic change: evidence shows that change has always been a feature of the Earth's climate. Apart from the evidence from the *Pleistocene Ice Age*, recent research has revealed the existence of a whole series of climatic trends on a variety of timescales. In North Africa, for example, there is evidence to suggest that within the last 20,000 years there have been periods when the climate was much wetter than it is today. Some of the major causes of climatic change that have been suggested include orbital fluctuations, sunspot activity, major volcanic activity, shifts in the broad pattern of oceanic circulation (*ocean currents*), and variations in the level of atmospheric *carbon dioxide*. At the present there is great concern that human activity has the potential to bring about climatic change through *global warming*, or the *greenhouse effect*.

climax vegetation: as the vegetation of an area develops and changes naturally through time, the characteristics and species of plants will alter until they reach a balance with the environmental conditions (soil and climate). This state is known as the climax community and will not change as long as the environmental conditions remain unchanged. (See also *succession.*)

clouds are visible masses of water droplets and/or ice crystals in the atmosphere. They are formed when air cools to *dew point* and vapor condenses. They generally develop when air is forced to rise either because of a relief barrier, where air masses are converging at a front, or as a result of convection. Clouds are classified into four types, based on shape and height: cirrus (wispy, hair-like), stratus (layer), cumulus (heaped) and nimbus (rain-bearing). The prefix alto- is used to indicate middle level clouds.

cloud seeding involves the introduction into clouds of condensation nucleii, salt particles or water droplets in order to induce greater precipitation. Solid carbon dioxide pellets, or dry ice, and silver iodide smoke can promote cloud growth and increase precipitation by triggering off the freezing stage of the *Bergeron–Findeisen process*, but the results of seeding are not predictable or reliable. Although modest increases have been reported, several experiments have resulted in a decrease in rain.

Club of Rome: an international team of economists, scientists, social scientists, civil servants and philosophers drawn from noncommunist countries. It was formed in 1968 to consider possible solutions to major world problems such as the gap between North and South, the pollution of the environment, unemployment and inflation. Their first considerations were published in *The Limits to Growth* in 1972. They predicted that if present trends in population growth, industrialization, food output, pollution and resource depletion were maintained, the limits to growth on the planet will be reached within the next 100 years, but they suggested that the trends could be altered to establish stability.

coastal landforms can be divided into two main groups, erosional and depositional. However, not all landforms can be explained in terms of present-day processes and therefore a third group reflects the effects of *sea level change*. Erosional landforms include cliffs, wave cut platforms, coves and features produced by the retreat of headlands such as pinnacles, stacks and arches. The main depositional landforms are beaches, spits and bars, but this group also includes features such as sand dune systems and salt marshes. Sea level changes may be *eustatic*, i.e. general changes that affect all oceans equally, or they may be *isostatic*, i.e. localized changes resulting from postglacial uplift of land. When sea level rises, lower parts of river valleys may be flooded to form *rias* or cliffs may be subjected to renewed attack. When sea level falls, raised *beaches* and raised platforms may result where features produced at sea level are abandoned. Former cliff lines may become degraded as they are now removed from the constant undercutting by marine processes and *subaerial weathering* reduces the angle of the cliff face.

coastal management involves the use of strategies to combat the effects of erosion, flooding and cliff falls. One group of responses can be described as the "hard engineering approach," i.e. constructing defenses. Sea walls are designed to absorb wave energy and to protect the cliff base from erosion; methods include bull nose concrete walls, stone and concrete blocks, and revetments. Groins aim to trap and stabilize sand and shingle by interfering with *longshore drift*. This encourages the formation of a beach as a means of absorbing wave energy. "Soft engineering approaches" or nonstructural responses try to maximize the natural process of development. *Beach nourishment* involves the replacement of material lost by longshore drift. Beaches are replenished with sand brought in from other parts of the coast. Beaches may also be stabilized by revegetating and reducing the slope angle. There is, however, another view that suggests that we should not fight nature and that it might be cheaper to let nature take its course and provide compensation to those people affected instead. Sea defenses need constant attention and this costs a lot of taxpayers' money. A more radical approach would be to prohibit development in coastal areas; such policies of "coastal retreat" have been introduced in parts of the USA.

coastal processes include the direct action of *constructive* and *destructive waves* and their influence on erosion, transportation and deposition of sediment. The characteristics of waves – height, velocity, wave length and wave period – are affected by wind speed and fetch. Waves erode material through hydraulic pressure, when air is trapped and compressed in a joint in a cliff and the increase in pressure weakens the rock. Abrasion also occurs when previously eroded debris is thrown against the cliff. In some coasts corrosion or solution can take place when minerals are dissolved. Material is transported by longshore drift and deposition of this sediment occurs in sheltered areas where there is low wave energy, for example in a bay or where the coastline changes direction. Coasts may also be attacked by nonmarine processes. *Subaerial weathering* by water, wind and frost can lead to forms of *mass movement* such as slumping, landslides and soil creep, which can remove material from the cliff face.

cold front: the boundary between a warm and cold air mass where the cold air undercuts the warm air, causing it to rise. The gradient of the cold front is steeper than that of the *warm front* and produces a more rapid uplift of air, giving rapid cooling and condensation. This is called an ana-cold front and produces cumulonimbus clouds and heavy but short-lived rain showers, sometimes accompanied by hail and sleet. Due to friction with the ground, the cold front may slow down at ground level, creating an overhang producing turbulent and squally conditions. The passage of a cold front is marked by a change in wind direction from southwest to northwest and a drop in temperature. When the warm air in the warm sector of the depression is subsiding (kata-cold front), conditions are more stable, producing stratocumulus cloud and light precipitation. The weather changes are more gradual than with ana-cold fronts.

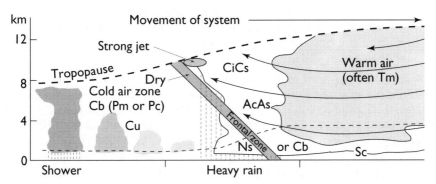

Cold front

collective farming was a type of agricultural system first developed in Russia after the 1917 revolution and practiced in other former communist states. The land was owned by the state and the collective leased to a large group of workers who operated it as a single farm holding, sharing its profits. Collective farms were intended to be self-governing but government intervention often set quotas and production targets.

colloids are very fine grained substances consisting of molecules dispersed in a liquid or gas. Clay colloids are finely divided clays dispersed in water, which when combined with humus form the clay *humus* complex in the soil.

colonizer plants are the first plants to move into an area as part of the *vegetation succession*. These species form part of the *pioneer community*, which is the first stage (*sere*) in a sequence that ends with the climatic climax community. They gradually modify the microclimate and soil, enabling other species to migrate in as conditions change. In a *lithosere*, a primary succession formed on bare rock, the initial colonizers are bacteria and single cell plants, together with mosses and lichens.

command economies: economic systems controlled by decision-making at the central government level. The former USSR and the Eastern European bloc with countries such as East Germany and Hungary operated command economies before their collapse in the 1980s. In each system almost the whole economy was planned, with targets being set for every point of production. The systems sought to distribute resources fairly among their citizens, but became highly bureaucratic and inefficient. They also became less able to supply goods of the right quality and quantity to meet demand. They were characterized by lower living standards than Western countries, which operate a *market economy*.

commercial farming is when the production is intended for sale in markets. Commercial farmers seek to maximize yields so as to maximize their profits.

comminution is the reduction in size of particles caused by their movement in rivers, glaciers, seas and in the wind. They may be reduced in size by striking one another, or by striking other objects as they are moved.

Common Agricultural Policy (CAP) is the scheme by which agricultural production within the *European Union (EU)* is organized. It was established by the Treaty of Rome with a number of aims:

- to increase agricultural productivity within member states
- to ensure a fair standard of living for their farmers
- to stabilize agricultural markets within and between the member states
- to ensure reasonable consumer prices
- to maintain jobs in agricultural areas.

These aims have replaced existing national policies and have often caused conflict between member states. Farmers are given guaranteed prices for their produce, and therefore have produced as much as possible. This has created surpluses in a range of agricultural products, known as "mountains and lakes." 70% of the EU's budget is spent supporting farming, yet farming only provides 5% of the EU's total income. The net gainers from the CAP are those countries with inefficient farmers, such as France and the countries of Southern Europe. The net losers are those with an efficient farming sector, such as the United Kingdom.

Supporters point to the facts that the EU is now largely self-sufficient in food, that more marginal farmers are still in business, and that farming is much more productive. Those against the CAP state that its bureaucracy is highly inefficient and open to corruption, and that it has caused food prices to be higher than they should be on a world scale.

communication systems are mechanisms by which information, goods and people are exchanged or moved from one area to another. Examples of communication systems include:

- telephone, facsimile and the Internet for information
- road, rail, air and water transportation systems for goods and people.

commuting is the daily movement of people from their place of residence to their place of work, and back again. There are two types of commuting:

- rural to urban – where the person lives in a small town or village and travels to work in a larger town or city
- intraurban – where a person lives in one part of a town or city and travels to work in another part of the same town or city. This often involves movement between the outer suburbs and the center, but also includes movement between inner city housing areas and edge-of-town industrial areas.

comparative advantage: the principle that countries can benefit from specializing in the production of goods at which they are relatively more efficient or skilled. In this way, the consumers within each country gain the maximum benefit from international trade.

Worked example

	Cost in days' work	
	Country A	Country B
To produce 1 unit of food	1	3
To produce 1 unit of clothing	2	1

If Country A specializes in food and Country B specializes in clothing, then both can have higher living standards. If 1 unit of food exchanges for 1 unit of clothing then trade allows Country A to obtain 1 unit of food and 1 unit of clothing in 2 days' work (instead of 3) by producing 1 unit of food for itself and another unit of food to exchange for 1 unit of clothing. Similarly, Country B saves 2 days' work (2 instead of 4).

competence: the diameter of the largest particle that a river can carry for a given velocity. Research has shown that a river's competence increases according to the sixth power of the velocity of that river, i.e. if a river's velocity doubles, then its competence increases by 64 times (2^6). This is because fast-flowing rivers have greater turbulence and are therefore better able to lift particles from the stream bed.

competition exists when more than one company or organization has an opportunity to meet the demands for a service or good. In this situation there is no *monopoly*. When competition is high, consumers usually benefit in the short term due to falling prices. However, in the longer term, one of the companies may not survive and its subsequent closure will result in less choice for the consumer. In some industries and services, competition may lead to wasteful duplication.

composition of a vegetation type refers to the *species* that make it up. When a vegetation is made up of only one species (for example, a pine plantation) it is called a plant society. However, vegetation is usually made up of a collection of species and is called a plant community. Tropical rain forests, where there are warm and moist conditions, have several thousand species of plants.

concentric urban model: this model was devised by Ernest Watson *Burgess* in 1923 in an attempt to explain the pattern of social areas within the city of Chicago, but it was later seen to have a wider application. The model is based on the ideas that both the growth of a city and the socioeconomic groupings of people spread outward from its central area to form a series of concentric zones.

At the center of the city is the *Central Business District (CBD)*, the focus of commercial, social and civic life. It is surrounded by the *transition zone* containing industrial premises, obsolete housing and slum property occupied by lower social groups of people and a high proportion of immigrants. This is surrounded in turn by a zone of working-class housing, occupied largely by people who have migrated out from the transition zone, but still need to live close to their place of work. Second generation immigrants also form a significant proportion of the population of this zone. The next zone moving outward is the zone of middle-class housing consisting of single family dwellings interspersed with exclusive residences and luxury apartment buildings. Finally, at the fringe of the urban area is the commuter zone. This is separated from the continuously built-up area by a *green belt*, but includes villages that are changing their character and function to become *dormitory settlements* for commuters who travel to work in the city each day.

Burgess' model has been widely criticized. For example, it makes insufficient reference to the siting of industry, which rarely forms a concentric zone anyway. Also, it does not account for the effects of both topographical features and transport systems.

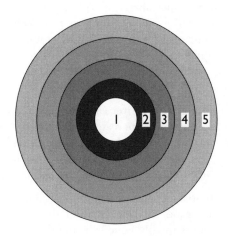

1 = CBD
2 = Zone of transition (inner city)
3 = Zone of working-class homes
4 = Residential zone
5 = Commuter zone

Burgess' model of urban land-use

condensation is the process by which droplets of water or ice are formed when water vapor is cooled to *dew point*. During the process, latent heat of condensation is released, causing a lowering of the *adiabatic lapse rate* in rising air. In the *atmosphere*, hygroscopic nucleii, such as salt particles, may attract water and cause condensation before the air is cooled to dew point. This produces cloud, fog and mist.

confidence levels indicate the degree of confidence that can be placed on the results of a statistical exercise. These are stated as the 68%, 95% and 99% confidence level. They indicate the level of probability that the result obtained is the correct one. For example, if after *sampling* the land use in an area, 40% of land is estimated to be woodland, and there is a sampling error of 2%, we can be 68% confident that the actual amount of woodland lies within ±1 sample error of the estimate, i.e. the actual amount of woodland lies between 38% and 42%. To reach the 95% confidence level we extend the range to ±2 sampling errors and for 99% confidence we extend to ±3 sampling errors. So, in this example, we can be 99% confident that the actual amount of woodland lies between 34% and 46%. In statistical tests 95% is an acceptable level of confidence.

conflict: this may result from opposing views over the ways in which a *resource* might be developed. Different individuals or groups may have different attitudes toward the exploitation of a specific resource. For example, the intensification of agriculture to increase production could lead to conflicts; the use of fertilizers may increase nitrate pollution of groundwater and streams, which will arouse opposition from the water supply industry and fishing interests.

coniferous woodland: in addition to the extensive areas of coniferous (*boreal*) forests of the high latitudes, large areas of woodland have been created by *afforestation*, mainly for the purpose of commercial timber production. These man-made woodlands are predominantly coniferous softwoods that produce timber with more varied uses than deciduous hardwoods and, in similar environmental conditions, softwoods grow more quickly.

connectivity is the extent to which the points in a network are interconnected. Actual networks are converted into topological maps, or graphs, in which the vertices (nodes) represent the points in the network and the edges (arcs) represent the direct links between the points. Connectivity measures the efficiency of the network; the more direct the links between a given number of points the more efficient the network. A number of measures such as the alpha, beta and gamma indices, the cyclomatic number and the Shimbel index can be used to assess and compare networks.

conservation is the protection, and possible enhancement, of natural and man-made landscapes for future use. In urban areas individual buildings or areas of settlement may be protected because of their historic interest; in rural areas species, habitats and landscapes may be protected. Conservation also encourages sensible use of resources and a reduction in the rate of consumption of *nonrenewable resources* as a means to achieving *sustainable development*. Efficient use of resources includes the adoption of less wasteful extraction methods, more efficient use of energy in processing and recycling of waste material.

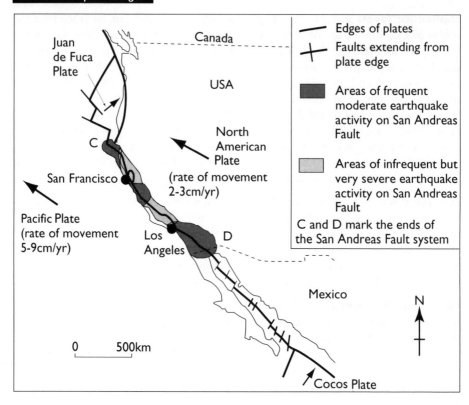

Conservative plate margin

conservative plate margin: in *plate tectonics* when two crustal plates slide past each other the movement of the plates is parallel to the plate margin.

The San Andreas Fault in California forms a boundary between the Pacific Plate and the North American Plate. Both plates are moving toward the northwest but the faster rate of movement in the Pacific Plate creates the impression that they are moving in opposite directions. The stresses caused by the movement of sections of crust past each other can trigger *earthquakes.*

constructive plate margin: in *plate tectonics* when two plates are moving away from each other magma flows upward and spreads, creating new areas of crustal material. In ocean areas this produces ocean ridges such as the Mid-Atlantic Ridge, and in continental areas it results in *rifts* such as the East African Rift Valley. They are also known as divergent margins.

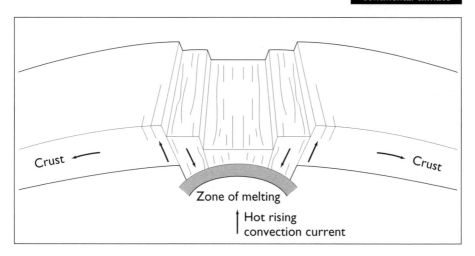

Constructive plate margin

constructive waves build up material on the beach and contribute to the formation of beach ridges and berms. They are waves of relatively low frequency, usually between 6 and 8 per minute, and low height. When the wave breaks there is a strong *swash* carrying material up the beach. The *backwash*, moving material down the beach, is much weaker and therefore less capable of removing sand and shingle. These tend to occur on low angle beaches where energy is dissipated over the beach. Material is constantly moved up the beach, increasing the gradient of the beach profile.

contagious: a disease that is capable of being passed on by direct contact with a diseased individual or by handling that person's clothing.

containerization was an innovation in the handling and transportation of cargo by ship in which goods are packed into containers of specified sizes at factories, taken by train and truck to a container port to be loaded on to specialized container ships. The international specified sizes for containers are 3, 6 or 12 meters long, 2.5 meters wide and 2, 2.5 or 3 meters high. Their introduction resulted in a large reduction in the traditional dock labor force, causing industrial unrest in some countries. The advantages of containerization are:

- the turn-around time (to load and unload cargo) for ships is considerably reduced
- with standard containers, standard handling facilities can be developed at ports
- easy storage of freight on dockside
- greater security for cargo, once containers have been sealed
- reduction in labor force required at docks.

continental climate is a general term covering the climate of those areas protected from or unaffected by maritime influence. It is most marked in the temperate to high latitudes of the northern hemisphere. In these regions the climate is generally characterized by:

- warm/hot summers with cold/very cold winters
- large annual ranges of temperature
- low levels of precipitation
- summer maximum of precipitation.

continental drift: the hypothesis that the present distribution of the continents is the result of the break-up of larger land masses followed by their drifting apart over long periods of geological time. The theory was put forward by Alfred Lothar *Wegener*, who suggested that the Earth's continents had once been joined together as one large land mass, which he named *Pangaea*. This "super-continent" had two main components, *Laurasia* to the north and *Gondwanaland* to the south. The evidence for such movements put forward by Wegener was:

- the general jigsaw fit of today's continents
- geological evidence of rocks of similar type, age and formation that occur in South America and Africa and also between the eastern part of North America and Europe
- fossil and *paleoclimatic* evidence that similar animals were once found in now widely separated areas along with similar climates.

Wegener's ideas were initially viewed with a great deal of skepticism as he was unable to suggest a mechanism for the movement. Later research, however, particularly in the field of *paleomagnetism*, has led on to the now widely accepted theories of *plate tectonics*.

continental shelf is the relatively shallow belt of sea bottom bordering a continental mass, the outer edge of which (continental slope) sinks rapidly to the ocean floor. It generally extends to a depth of about 100 fathoms or 200 meters. In recent years its definition and delimitation have assumed increasing importance in international law in connection with the ownership of minerals, particularly oil and gas, that lie beneath it.

contract farming is when large *agribusiness* companies, increasingly involved in the *food chain*, offer contracts to farmers to supply them with produce. Contract farming is particularly found with products such as sugar beet, vegetables, fruit, and pig and poultry products. Large *transnational* firms are often involved, e.g. Unilever and Nestlé.

contracting out means placing with independent suppliers a task that used to be carried out inside an organization. *Private sector* firms, local government or nationalized industries may contract out services such as cleaning, refuse disposal, or even the production of components. Contracting out is a reversal of *vertical integration*.

Pros:
- may lead to lower costs as the contractor's wage rates do not have to be as high as those within the organization
- putting the service out to tender invites new management thought on how to improve efficiency

Cons:
- the subcontractor's employees may be less motivated to provide the quality the organization wants
- from the employees' viewpoint, working for a subcontractor may mean more intensive work for less pay.

conurbation: a large and almost continuous *urban area* built up from separate centers that through urban growth and sprawl have joined.

convection occurs when the lower air heated from the ground expands and rises. As long as the rising air is warmer than its surrounds, it will continue to rise. This creates *instability* within the atmosphere and cumulo-nimbus *cloud* formation. Convection can also occur within water bodies.

convection currents are movements within the *asthenosphere* that move the crustal *plates.*

convergence is when air flows into an area from different directions. When this happens at or near the surface, the air is forced to rise. In the tropics, the *inter-tropical convergence zone (ITCZ)* is an area of surface convergence.

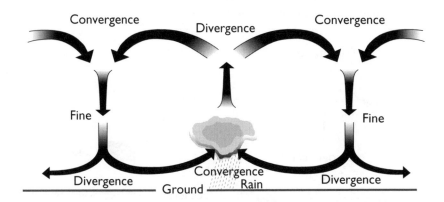

Convergence/divergence

cooperative agriculture: individual farmers who are not in a good position to meet market demand, particularly that of major food retailers, can improve their position by joining with other farmers in a producer cooperative. The advantages to the farmers are:

- they can negotiate favorable contract prices with retailers
- they can provide a continuous supply of uniform quality produce
- processing, grading and packaging costs are shared
- farm inputs can be obtained in bulk.

coral reef is an accumulation of coral formed around the edges of landmasses such as the eastern coast of Australia and around the shores of islands. Corals are tiny animals (polyps), related to sea anemones, that exist in large colonies. As they die they leave behind a hard skeleton consisting of calcium carbonate, which looks like rock. The accumulations of such deposits create coral reefs with living coral on the top.

The growth of coral takes place in sea water with temperatures not less than 21°C, and only within 30–40 m of the sea surface. They need clear oxygenated water with plentiful supplies of microscopic zooplankton upon which to feed. They cannot live in fresh or silt-laden water.

The formation of coral reefs and atolls is still very much a source of debate. Charles Darwin put forward a theory that still has some credence today. He imagined that a volcanic island would have been created in the middle of the ocean (a *hot spot*). Coral would have grown around the fringes of this island, but with time the island may have subsided or sea level may have risen, creating a rising reef of coral around it (so long as the rate of land subsidence/sea level rise was not greater than the rate of coral growth). The result of such movements would cause first a fringing reef, then a barrier reef, and then possibly an atoll.

Coral is under threat in a number of areas of the world. As coastal *mangrove* swamps are cleared, their ability to trap sediments that cleanse coastal waters has reduced, and the offshore polyps are being smothered by dirty water. Tourists are also creating problems through powered water craft increasing turbidity and toxicity, and by souvenir hunting. A natural predator, the Crown of Thorns jellyfish, is also on the increase, although the reasons for this are less clear.

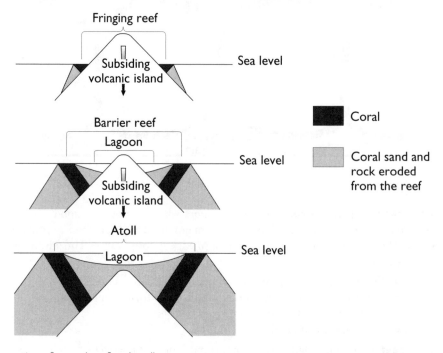

Formation of a coral reef and atoll

core is the name given to the interior of the Earth. There are thought to be two parts to the core – the inner core, which has properties similar to that of a solid, and the outer core, which has properties similar to that of a liquid. It is thought that the core is very dense, and that it is probably composed of iron, with lesser amounts of elements such as nickel. The core also has very high temperatures (5500°C) and very high pressures.

Core also refers to an area of concentration of economic development, associated with the models of *Myrdal*, *Hirschman* and *Friedmann*.

Coriolis Force is the effect of the Earth's rotation on air flow. In the northern hemisphere, the Coriolis Force causes a deflection in the movement of air to the right, whereas in the southern hemisphere it is to the left.

corrasion an alternative name for *abrasion.*

correlation is the degree of association between two sets of data. This involves the comparison of one set of variables with another. It can be done in two main ways:

- the calculation of a correlation coefficient that summarizes the relationship between the two sets of variables. The *Spearman Rank correlation* coefficient technique is one of the most useful. It is based on the ranks of the individual values of the two variables rather than the values themselves. All such calculations should also be tested for their statistical *significance*
- the construction of a *scatter graph* with the independent variable on the x axis, and the dependent variable on the y axis.

A positive correlation indicates that as one variable increases, so does the other. A negative correlation indicates that as the independent variable increases, the dependent variable decreases. It is also important to note that a high correlation between two sets of data does not necessarily prove that there is a causal relationship between the variables. It cannot be assumed that a change in variable A will cause a change in variable B. Further investigation may be required.

corrie: see *cirque.*

cost-benefit analysis means evaluating the financial and social costs of a course of action against the financial and social benefits. For this kind of evaluation, an estimate of the external costs (such as environmental damage) must be made along with the benefits (such as increases in employment). This is often difficult to calculate. A person in employment can save the government a known average amount, but it is more difficult to quantify the damage to the environment.

A cost-benefit analysis of a nuclear power plant, for example, would consider the possible damage to the countryside (both in construction and a possible nuclear accident) against the cost of the power produced. The analysis would also have to look at competitors in electricity generation and the cost-benefit, for example, of the closure of a coal-fired station and its effect on the coal mining industry.

cost of living: the amount of money spent by the average household over a period of time. This is difficult to evaluate precisely because people have differing spending priorities. Younger people with children, for example, spend a high proportion of their income on essentials such as food and housing, whereas older people may spend a lot more on things like entertainment and holidays. Changes in housing costs will affect the younger people's cost of living more than the older people. The government attempts to measure the cost of living using a "basket of goods" that research has identified as the average household's expenditure. The best-known measure for the cost of living is the retail price index (RPI).

cottage industry is an industry in which employees work in their own homes, often using their own equipment.

Council of Ministers: the *European Union*'s main decision-making body. Its membership comprises one government minister from each of the member states.

The minister chosen will depend upon the issue under discussion.

counterurbanization: the process of population decentralization as people move from major urban areas such as *conurbations* to much smaller urban settlements and rural areas. This trend was first noted in the USA in the early 1970s, but has spread to most industrialized countries. The explanation for this movement is to be found in a combination of reasons, but above everything else, counterurbanization seems to be a reaction against large cities. There is also the rise of the new technology, particularly in the area of communications, which seems to be reducing the need for people and activities to *agglomerate* in towns and cities.

craters are the holes at the top of a volcanic cone. They are usually circular depressions surrounded by low rims of ejected debris, and they may become occupied by a lake. Craters can vary in size from hundreds of meters in width to several kilometers.

cratons (also known as shield areas) are the oldest (over 550 million years old) and most stable interior parts of continents. They are located away from tectonic plate margins, and form the nucleus around which new mountain belts have developed. They lack active volcanoes or earthquake activity, and are severely eroded, forming extensive areas of low plateau. Examples include the Canadian Shield, the Baltic Shield and the Brazilian Shield.

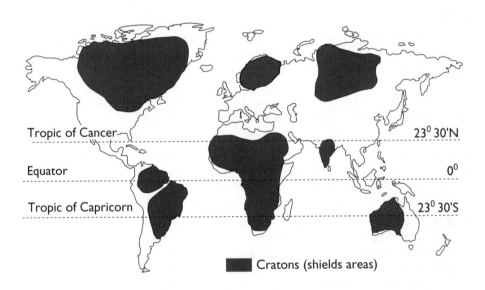

Distribution of cratons

creep is the slow downhill movement of soil and other material such as *scree* (talus). It operates on slopes steeper than 6°. A number of processes, each of which are capable of producing only slight movements, combine to cause it:

- raindrop impact – the beat of raindrops onto a loose surface during a period of intense rainfall
- the expansion and contraction of the soil caused by either seasonal wetting and drying (especially clays), or seasonal and diurnal freezing and

thawing of the soil surface. In both cases, expansion causes the surface of the slope to heave at right angles to the original surface, but it then falls back under gravity in a more perpendicular manner

- the swaying of vegetation during windy spells
- the treading of both wild and domesticated animals.

The rates of creep vary in differing environments. In temperate humid environments rates are commonly 1–2 mm a year, but in moist tropical areas the rate can be 3–6 mm a year. Common forms of evidence of creep are tilted telephone poles, terracettes, and the build-up of soil on one side of a stone wall causing the wall to lean to one side.

cross-profile is the view of a feature or landform from one side to another. For example, a river valley in an upland area has a typical V-shaped cross-profile with steep sides and a narrow bottom. A glaciated valley has a U-shaped cross-profile, with steep sides and a wide flat bottom.

crust is the name given to the outer layer of the Earth, generally thought to be between 6 and 70 km thick. It is separated from the underlying *mantle* by the *Mohorovicic* discontinuity (named after a Yugoslavian seismologist). There are two types of crust:

- continental crust (also known as the *sial*) – with a lighter average density (2.7 gm per cubic cm), consisting largely of granitic rocks
- oceanic crust (also known as the *sima*) – with a relatively higher density (3.0–3.3 gm per cubic cm), consisting of basaltic rocks.

The oceanic crust is continuous around the planet's surface, whereas the continental crust only occurs where there are continental land masses, and it rests on top of the basaltic layer.

The continental crust is also much older than the oceanic crust. In Greenland the continental crust is more than 3500 million years old, whereas the oceanic crust is at no point more than 250 million years old.

cuesta: a landform with a steep scarp slope and a gentle dip slope, which forms on sedimentary rocks that are gently dipping. They are formed by differential erosion

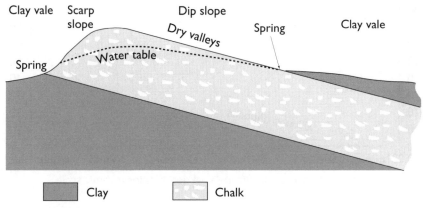

Scarp and dip slope escarpment (cuesta)

of alternating resistant and less resistant strata. Chalk and limestone, because of their permeability, are left as upstanding escarpments while weaker sands and impermeable clays are subjected to greater erosion and produce lower lying vales.

cumulative causation is a process suggested by Gunnar *Myrdal* in 1957 to explain regional differences in economic growth. When a region has some form of initial advantage such as raw material resources, nearness to a port or is the source of some innovative industrial invention, this may set cumulative causation in motion. The establishment of a leading industry may trigger off other developments; other industries supplying that industry with inputs may be attracted to the region. These are termed ancillary industries. Subsidiary firms, which use outputs or products of the initial industry, may also be attracted. These industrial *linkages* can lead to the agglomeration of industry and other *agglomeration economies*. This activity may lead to other *multiplier effects* with in-migration of labor and increased employment in construction, transportation and services to supply the needs of the larger population. Economic growth leads to a higher local tax yield and improved infrastructure, which makes the region more attractive to other industries; in this way one development leads to other growth and the process is cumulative. The process can work in reverse, where the closure of a major employer leads to a downward spiral in the economy of a region.

cumulative frequency: the data for each class in a distribution is converted into a percentage and each percentage figure is added successively. For example, if 5% of the population is under 5 years, 4% is 5–9, 6% is 10–14, then cumulatively 5% are under 5 years, 5 + 4 = 9% are under 10, and 5 + 4 + 6 = 15% are under 15. This would be continued until all groups are included when the cumulative total percentage would be 100%. It can be displayed on a cumulative frequency curve with cumulative frequency on the vertical axis (scale 0–100%) and range of values on the horizontal axis. This would allow identification of the *median* (50th percentile) and upper and lower *quartile* positions (75th and 25th percentile).

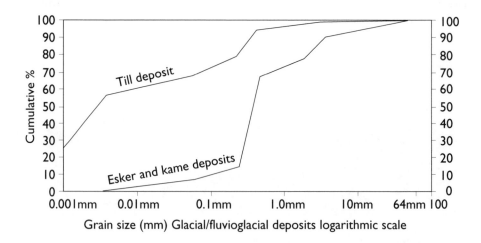

Cumulative frequency graph: grain size distribution in glacial/fluvioglacial deposits

In the accompanying graph the till contains a wide variety of grain sizes. Esker and Kame deposits contain a small percentage of smaller grains and a large percentage of grains between 0.3 mm and 5 mm.

cusps are small arcuate hollows or embayments that form on beaches. They vary in size up to 50 meters across and lie parallel to the water's edge with projecting low ridges pointing toward the sea. The sides of the cusp channel incoming swash into the center of the embayment, producing stronger backwash in the central area, which drags material down the beach and deepens the embayment. The low projecting ridges contain coarser sediment than the hollows.

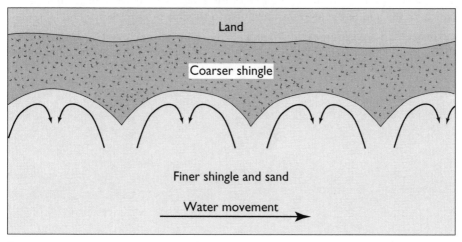

Cusps

cycle of poverty: a vicious circle that exists within the economy of rural regions of Third World countries. Farmers who lack money cannot invest in better seeds,

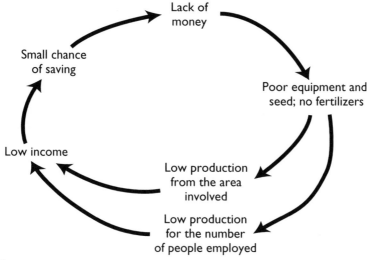

Cycle of poverty

equipment and fertilizers. This inevitably results in poor crop yields, giving low incomes, which in turn relate back to the lack of investment in the system. Loans and grants from government or international funds will help to break the cycle as long as the farmers are also educated in more efficient agricultural methods. Moreover, helping farmers to set up *cooperatives*, where resources are pooled, will help individuals in breaking the cycle.

cyclogenesis is the sequence of events leading to the formation of cyclones, especially the middle latitude depressions. Cyclogenesis is caused by the convergence of *air masses* at ground level due to divergence of air in the upper *troposphere*. The main areas of cyclogenesis are along the line of the Polar Front in the North Atlantic and North Pacific and in the Mediterranean Basin (mainly in winter).

cyclone: the name given to the low-pressure systems accompanied by severe weather that occur in the western Pacific Ocean and affect areas such as the Philippines, Taiwan, Hong Kong, China and Japan. (See *hurricane*.)

Dalmatian coast: a drowned coastline where the main relief trends run parallel with the line of the coast. The ridges of upland produce elongated islands separated from the mainland by the flooded valley areas. The name originates from the Adriatic coast of Dalmatia. Alternative names for this type of coastline are concordant and Pacific-type coasts.

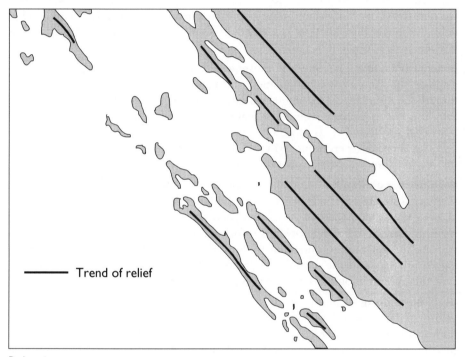

Dalmatian coast

DALR: see *dry adiabatic lapse rate.*

death rate: the number of deaths in a year per 1000 of the population. This is referred to as a crude death rate as it takes no account of the differences in the sex/age structure of populations.

debris is the name given to the fragments of loose material that are produced as a result of the breakdown of minerals and rocks. This material may accumulate on slopes as *scree* or *talus,* or may be transported by mass movements, *channel processes,* wind action or as *supraglacial, englacial* or *subglacial material.*

debt (Third World): there are two main types of international debt: commercial debt, which is owed mainly by government departments, public corporations and private companies; and official debt, which is the repayment and accumulated interest due on loans through aid programs. Although some of this aid comes as grants or gifts, much of it is in the form of loans that have to be repaid. During the 1970s increasing amounts of foreign money poured into developing countries, much of it from commercial banking in industrial countries. Governments and their lending agencies were also providing money for aid projects set up with long repayment periods in anticipation of a good return as economies began to expand in developing countries. These recipient countries assumed they would be able to meet their repayments and this confidence led to a rapid increase in debt levels, rising from $68 billion in 1970 to $686 billion by 1984. As debt rose so did annual interest charges. The recession in industrial countries in the 1980s led to a fall in trade and commodity prices and an increase in interest rates. Many countries were using new loans and aid to service existing interest payments on previous debts. The debt crisis became public in 1982 when Mexico announced that it could not pay its foreign debt. Although the situation has eased in some countries, it remains a serious obstacle to development.

decentralization: the outward movement from established central areas, for example the movement of population and employment from the inner city areas toward the suburbs or to smaller urban centers. This trend may be a natural response to the negative aspects of higher crime, noise, pollution and high land costs that are found in central locations, or the process may be encouraged by agencies such as government trying to spread investment and development from the core area toward the periphery.

deciduous woodland: a woodland comprised of trees that are mainly broad-leaved such as oak, birch, ash and beech and that shed their leaves in the fall before the onset of lower winter temperatures. As soil temperature falls the tree roots can only absorb small amounts of water and growth is retarded. Leaf loss reduces *transpiration* and the demand for water. In winter there is also a decrease in the water content of cells, together with a rise in the sugar concentration of the sap. This allows water to enter the cells but prevents its outward passage thus preventing excessive loss of moisture.

decision-making is the process in which alternative strategies toward achieving a goal or different solutions to a problem are evaluated and a decision is taken. A number of theoretical models have used the concept of *economic man,* an optimizer who aims to maximize returns and who has perfect knowledge enabling rational decisions to be taken. In behavioral geography it is recognized that in the real world, decision-making is likely to be influenced by a number of factors such as decision-makers' perceptions, circumstances, access to relevant information and their ability to interpret information and analyze different alternatives.

decomposition is the breakdown of plant and animal remains, which releases energy and nutrients into the soil. Organisms such as earthworms, mites and slugs help in the decay and incorporation of leaf litter in the soil, and fungi and bacteria secrete enzymes that break down organic compounds in this detritus. Organisms that are decomposers are called detritivores.

deflation is the removal of small grained particles by wind action. It is usually a small-scale process that involves the moving of previously loosened sand grains, leaving behind small hollows and blow outs.

deforestation is the deliberate clearance of forested land by cutting or burning. It can have a major impact on surface water flows, channel *hydrographs* and soil erosion as the interception layer of the canopy and the soil binding properties of the roots are removed.

deglaciation is the retreat of an ice margin, either the snout of a valley glacier or the edge of an ice sheet. It occurs when there is a negative mass balance, and *ablation* at the ice margin exceeds the rate of supply of material accumulating in other parts of the glacier or ice sheet. Large amounts of meltwater are released during deglaciation that can produce glacial lakes and overflow channels, and if the deglaciation is extensive *eustatic* changes of sea level occur. As the weight of the ice is removed during melting, the land mass undergoes *isostatic uplift* and readjustment.

deglomeration: the opposite of *agglomeration*, as firms disperse from a site because of increased costs such as those of labor, transportation and land. Firms may find expansion difficult in the area and there may also have been a decline in the local market.

deindustrialization occurs when there is an absolute, or perhaps relative, decline in the importance of manufacturing in the industrial economy of a country and a fall in the contribution of manufacturing to *GDP*. In spatial terms it is most severe in the areas that have traditional heavy industries, such as iron and steel, chemicals, shipbuilding and textiles, which in the USA have suffered from strong overseas competition from countries where new technology and less unionized labor practices have been adopted.

delta: a landform produced by the deposition of sediment at the mouth of a river as it enters a sea or lake.

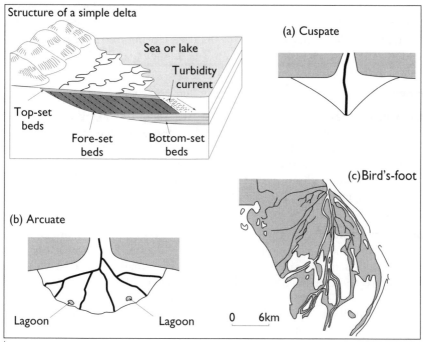

Delta

Deposition occurs when the river's velocity and sediment-carrying capacity decrease as it enters the lake or sea, and bedload and suspended material are dumped. In addition, clay particles *flocculate* due to the chemical change in sea water and settle on the bed. Deltas form where the rate of deposition exceeds the rate of sediment removal. Unless the sediment load is very large, as in the case of the Mississippi Delta, the feature develops in coastal areas that have a small tidal range and limited wave action or offshore currents. Deltas are classified according to their shape as arcuate, bird's-foot or cuspate.

demarcation is the dividing line between one job function and another. For example, it used to be common for maintenance workers in factories to be split into mechanics and electricians. Nowadays companies prefer multiskilling, which does away with demarcation lines and therefore prevents *demarcation disputes.*

demarcation disputes are industrial disputes about attempts to change existing *demarcation* lines. They occur because of the possible threat to employment if two jobs are combined into one. They were a particular feature of industries such as shipbuilding.

demesne: the land in a medieval manorial estate that was retained by the lord for his own use, or for use by his servants. Land outside the demesne was occupied by the villagers.

demographic transition model describes a sequence of changes over a period of time in the relationship between *birth* and *death rates* and the overall population change. The model suggests that all countries should eventually pass through similar stages, although the rate of change will vary from country to country. The model is a generalization based on observation, and particular countries can be expected to deviate from the pattern suggested; some may even miss out a stage. Some demographers believe that *developed countries* are moving into a new stage beyond that shown in the model where deaths will be higher than births, and population totals will begin to fall.

demography is the study of population numbers and change.

dendrochronology: the method of using tree rings (by taking a core through the trunk) to find the age of the tree and to use as evidence of past climates. Each year's growth is shown by a single ring, but when the year is warmer and wetter the ring will be larger, showing more growth than if it was cooler and drier. Research has shown, though, that precipitation is perhaps more important to growth than temperature. The oldest trees found by this method are over 5000 years in age.

dependency ratio: the relationship between the economically active or working population and the noneconomically active population. For ease of calculation, the active population is taken as all those in a population aged 16–65 and the nonactive are those aged under 16 and over 65. In the 16–65 age group, for the purposes of calculating the ratio it does not matter if a person is employed or unemployed. Ratios in the *developed world* usually lie between 50 and 75 but in *developing* countries the ratio is high, sometimes over 100. The ratio is calculated as follows:

FORMULA $\dfrac{\text{children } (0\text{--}15) + \text{elderly (over 65)}}{\text{all population } 16\text{--}65} \times 100$

Worked example: dependency ratio of the USA in 1990 (in thousands)

$$\frac{56,855 + 31,239}{160,571} \times 100 = 54.86$$

dependent variables are those that are directly affected by variations in another variable (the *independent variable*). For example, when considering the relationship between the amount of rainfall and altitude, rainfall varies with altitude, rather than altitude being a function of rainfall. Therefore, rainfall amount is the dependent variable – it "depends" on altitude. The dependent variable should be placed on the y axis of *scatter graphs* when attempting to identify a *correlation* between two sets of data.

deposition: the laying down of solid material, in the form of *sediment* such as mud and sand, on land, on the bed of a river or on the sea floor. Deposition is carried out by several agencies such as rivers, glaciers, wind and marine processes and usually follows from the *erosion* of the land and the transportation of the resulting debris. Deposition is one of the main ways in which the land is built up.

depression: an area of low atmospheric *pressure* with a roughly circular pattern of isobars that occurs in the mid-latitudes. Most depressions form along the zone of contact between cold, dense *polar* air and the warmer, lighter air from *tropical* latitudes. This zone of contact is known as the polar *front*. Theories of origin were first put forward in the 1920s by a group of Norwegian meteorologists, but with

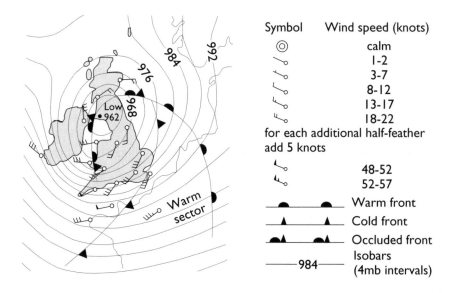

Depression

improved technology, particularly *satellite* imagery, refinements have been made to the original suggestions. (See also *cyclone* and *cyclogenesis.*)

deprivation: the measure of an individual's well-being when it falls below a level generally regarded as a reasonable minimum. To measure the extent of deprivation, geographers have used a range of indicators from the areas of employment, housing, health and education. In developed countries, the problem is seen to be one that particularly affects the central residential parts of urban areas, the *inner cities.* Some of the indicators that have been used to show deprivation within these areas include: unemployment, mean family income, overcrowding, levels of public housing, numbers of children on free school meals, electricity disconnections, *life expectancy,* levels of basic amenities (e.g. toilets, hot water).

deprivation cycle: a *downward spiral* that is particularly seen within *inner cities.* Such cycles prevent the poor from raising their living standards and contribute to the quality of life experienced within such areas.

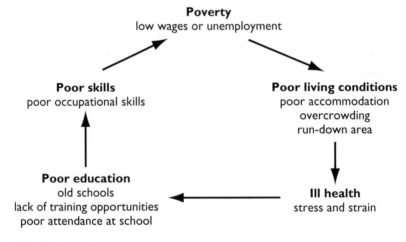

A typical deprivation cycle

deregulation is the removal of government rules, regulations and laws from the workings of business. This could include moves such as ending *monopoly* rights to supply services such as letter deliveries or bus services.

Pros: • fewer regulators need be employed therefore cutting public spending and theoretically, taxation
• encourages more competition

Cons: • rules were set up in the first place to give protection to workers; should they be removed?
• the effects of deregulation may be to provide only that which is profitable, and not what the public want, e.g. buses running only at peak periods but not at other times, such as late at night.

derelict land: previously used land that is now abandoned, such as old dock areas, factory buildings, areas where inner city housing has been cleared, waste heaps from mineral working. These areas are all in need of some form of reclamation.

desertification is the spread of desert-like conditions into neighboring semiarid regions of bush, grassland or woodland. Since the early 1970s there has been increased frequency of drought in the Sahel, causing severe famine in Sudan, Mali and Niger. Although low levels of rainfall, and unreliability of rainfall, do influence soil and land degradation that leads to desertification, drought problems have also been aggravated by human activity. Increased population pressure in relation to the carrying capacity of the resource base has led to overuse of the land through over-grazing and overcultivation. The scarcity of fuel supplies has also led to the wide-spread removal of woodland for firewood. This results in less interception of rainfall, reduced infiltration, faster run-off and greater soil erosion. Vegetation removal also exposes the soil to greater wind erosion.

desire line is a line on a map that represents the movement of people from their homes to certain destinations such as schools, offices, malls, holiday resorts and other such places. A desire line is drawn in a straight line between the home or home area and the destination, with no account being taken of the actual route. The thickness of the line is proportionate to the number of people who are moving. A suitable scale should be identified before the map is drawn so as to avoid excessive clustering at the destination point.

destructive plate margin: in *plate tectonics*, when two plates are converging or colliding this leads to the destruction of plate material. *Subduction* zones occur when

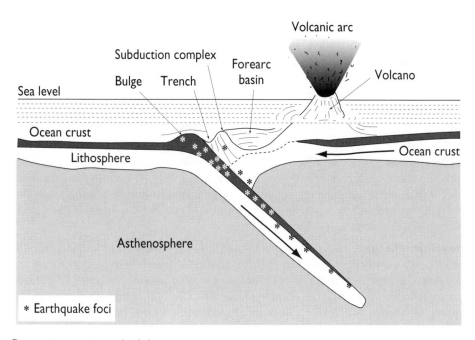

Destructive margin and subduction zone

an oceanic plate subducts beneath another oceanic or continental plate. When two oceanic plates converge subduction produces a deep ocean *trench* and an *island arc*, a line of volcanic islands. If the convergence involves an oceanic plate and a continental plate, the subduction produces a line of mountains along the edge of the overriding continental plate, as in the case of the Andes. These are mountains produced by the folding and faulting of sediments that accumulated on the continental shelf on the margin of the continental plate, and by the melting of the subducting oceanic lithosphere that produces rising volcanic magma. If subduction leads to the closing of an ocean so that two continental plates collide, complex fold and fault structures are produced – for example, the Himalayas. The two crustal plates are of similar density, which is much lower than the underlying *asthenosphere*; as a result there is little subduction and the collision causes buckling of the plates. These boundaries are also called convergent margins.

destructive waves transport sediment back down the beach, producing an overall loss of material. They are steep waves that break at a high frequency of 13–15 per minute. As the wave crest breaks, water plunges downward at a steep angle. This produces very little *swash*, which reduces the movement of sediment up the beach. The *backwash* is stronger and is more effective in dragging material down the beach.

developed is a term usually applied to a country or region that has a high standard of living and an advanced economy based on the effective utilization of resources. Such areas are usually more dependent on manufacturing and service industrial activity and have much lower levels of primary employment. They may be referred to as *economically more developed countries (EMDCs)*, which acknowledges that those countries that are not "developed" in the sense of this definition may be developed in other noneconomic ways, such as with regard to cultural, religious or social conditions.

developing is a term usually applied to a relatively poor country or region that has a low standard of living but is beginning to achieve some economic and social development. Such areas tend to be more dependent on primary activities and some are significant producers of minerals (Brazil) or oil (Nigeria). Some countries have developed industrial growth and are emerging as *newly industrialized countries (NIC)*.

development is the use of resources and the application of available technology to bring about an increase in the standard of living within a country. Earlier views on "development" emphasized economic expansion and increased output, but the current view is much broader, involving social and cultural advancement as well as technological change and economic growth.

development gap: the difference between the economic development of the *economically more developed countries (EMDCs)* of the *North* and the poorer countries of the *South*. To many, this gap is widening and is seen as a direct result of policies pursued by the richer countries. There are many ways in which this gap can be viewed when considering diet, education, health and housing, but perhaps the most revealing statistics are:

NORTH
- 25% of world's population
- 80% of world's income

SOUTH
- 75% of world's population
- 20% of world's income

development models describe the varying ways in which nations or regions have changed their use of resources, discovered new uses for them, and exploited their own human skills. They all concern the manner in which manufacturing and service industries have emerged from an agricultural base. There are number of development models, of which the main ones are:
- the *Clarke-Fisher model*
- the *Rostow model.*

dew is the deposition of water droplets onto the surface of grass and the leaves of plants. It forms under clear, calm, anticyclonic conditions when there is rapid heat loss at night. The *dew point* is reached as the air cools, and the moisture in the air condenses onto these surfaces.

dew point is the temperature at which a body of air at a given atmospheric pressure becomes fully saturated. If an unsaturated body of air is cooled, a critical temperature will be reached when its *relative humidity* becomes 100% (i.e. saturated). This temperature is the dew point, and further cooling results in *condensation* of excess water vapor.

diastrophism: a general term for the action of those movements that produce relative or absolute changes in the position, level or attitude of the rocks forming the Earth's crust. Diastrophism is usually classified in two groups:
- *orogenic* movements, or mountain building, which involves intense *folding, faulting* and thrusting, causing much deformation to the rock strata
- *epeirogenic* movements, which are less intense and may only involve uplift and at the most some gentle folding with associated faulting.

Some authorities include within diastrophism *isostatic* and *eustatic* movements and also that of molten rock (*igneous* movement).

diatoms: minute single-celled algae with hard shell-like skeletons composed of silica. These skeletons are distinctive for each diatom species and are preserved in the *sediments* on lake bottoms when the organism dies. They can, therefore, be used to assess the history of the lake's development when cores from the sediment are analyzed.

diet refers to the average food intake of people, and is usually measured in megajoules per capita per day. The United Nations regards 10.8 megajoules per capita per day as being appropriate for healthy living. However, the quality of diets varies within the world, in terms of both the amount of food and the quality of food consumed. A balanced diet should contain:
- proteins – in meat, milk, eggs – to build and renew body tissues
- carbohydrates – in cereals, sugar, fats, potatoes – to provide energy
- minerals and vitamins – in dairy produce, fruit and vegetables – to prevent many diseases.

People who do not eat enough food are said to be undernourished, whereas those who do not eat the right diet suffer from *malnutrition*. Both of these conditions may lead to a variety of diet deficiency diseases.

diffluence happens when a small *glacier* breaks away from a main glacier and crosses over a drainage divide by means of a low col. In some cases, this can mean that the smaller glacier is forced upward and over the divide.

diffusion is the process in which an *innovation* is gradually adopted by more and more people through time and across space. At first there are the initial innovators, the leaders, followed by the first adopters of the idea. With time more people will adopt the innovation and its use will expand across an area.

discharge is the volume of water in a river passing a measuring point in a given time. It is calculated by multiplying the velocity of the river by the cross-sectional area of the river at the measuring point. It is measured in cubic meters per second, or cumecs.

discrete variable: in statistics, this refers to a variable that can only take a particular whole number value, such as the number of cars in an urban area or the number of people in a country.

diseconomies of scale: factors causing higher costs per unit when the scale of output is greater, i.e. causes inefficiency in large organizations. The major causes relate to the increased costs of internal communications and decision-making within large companies, which may not be as efficient as those of a smaller company.

dispersion measure: in statistics, these are the methods of calculating or displaying the distribution of a set of numbers around the *central tendency* (*mean*, *median*). The main statistical methods are the *range*, the *interquartile range* and the *standard deviation*. Visually, data can be represented on a dispersion diagram (see below), *histogram*, frequency polygon and a *cumulative frequency curve*.

Worked example

Annual rainfall totals at two widely separated stations over a period of 25 years

Station A	71	93	63	74	71	82	65	79	55	86
	84	70	79	98	75	89	71	79	68	75
	81	83	61	74	84					

$\bar{x} = 76.4$ cm $\sigma = 9.9$

Station B	59	90	86	36	53	123	90	43	111	68
	74	79	99	58	73	38	120	60	68	77
	89	80	78	87	48					

$\bar{x} = 75.5$ cm $\sigma = 22.9$

$\bar{x}$ = mean

σ = standard deviation

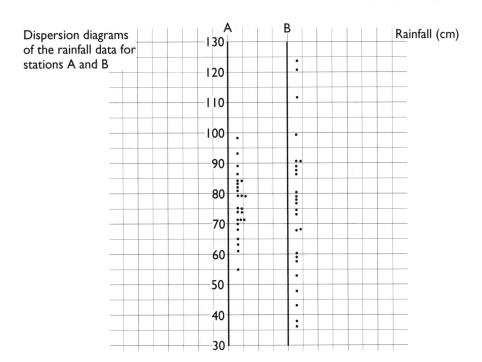

Dispersion diagram

distance decay: the fall in the amount of movement or spatial interaction between two centers, the greater the distance that they are apart. This is seen in the *gravity model* where flows between places are inverse to the distance that separates them, e.g. the number of people traveling to a mall to shop declines as the distance away from that mall increases.

distributary is a branch from a river that does not return to the main stream after leaving it. Distributaries are common in *deltas* such as those of the Mississippi and the Nile.

distribution: this can have two meanings:
- the pattern made by the occurrence of a feature within a given area, such as the distribution of public housing within a city
- the entire process of getting products from the producer to the consumer.

distribution channels: the stages of ownership that take place as a product moves from the manufacturer or producer to the consumer. (See *retailing* and *wholesaling*.)

Main channels of distribution

TRADITIONAL	MODERN	DIRECT
↓	↓	↓
Producer	Producer	Producer
↓	↓	↓
Wholesaler		
↓		
Retailer	Retailer	
↓	↓	↓
Consumer	Consumer	Consumer

Main channels of distribution

divergence is where air flows out of an area. Divergence near the ground results from/in air sinking (*subsidence*) and the creation of high pressure systems such as the subtropical high pressure zone (for diagrams see *convergence* and *atmospheric circulation model*).

diversification is the spreading of business risks by reducing dependence on one product or market. It is an important objective for companies of all scales, from small ice-cream manufacturers wanting security from cool summers to large *trans-nationals* wanting to move into new growth markets.

Pros: • improves prospects for long-term company survival
• can enable companies within saturated markets to find new growth opportunities
• provides new outlets for a company's skills and resources

Cons: • with expansion, companies may find that *diseconomies of scale* can arise
• companies may lack expertise within new markets, which can lead to disappointing results
• the company's core business could be weakened as resources are redirected toward new opportunities.

divide: the boundary that separates one *drainage basin* from another. It is also known as a watershed.

doldrums: the old name for the area of the *intertropical convergence zone* where there is a belt of generally light winds coupled with high temperatures and humidities. In the days before steam, sailing ships would often find themselves becalmed in these areas.

dormitory settlement refers to a rural village that has become increasingly urbanized in recent years, and is largely occupied by people who work in nearby towns and cities (commuters). It has new developments of houses or townhouses, frequently occupied by families or retired people. Most people own one or more cars, and shop as well as work in the nearby towns and cities. Consequently, the provision of services such as stores and public transportation is low. However, local schools are often enlarged.

doubling time is the number of years that it has taken for the world's population to double in size. It may also be used for the similar growth of population of a smaller area such as a continent or country. Doubling time has progressively become lower, which is evidence for the rapid population growth that has taken place over the last 200 years. The world's population is expected to double over the next 40 years.

downward spiral: when decline occurs within a region it may be irreversible, as the region loses its more motivated people, less investment is attracted, which in turn means that more people will leave, leading to even less investment and so on. Such a feature is also known as a *vicious circle.*

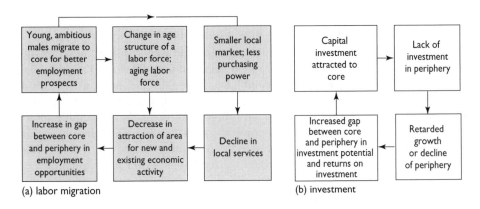

Downward spiral typical of the periphery in a developing country

draa: an extensive desert dune often having smaller dunes on the sand surface. They may be produced by the amalgamation of a series of smaller dunes. These dunes can extend over 5000 meters in length and reach heights of 400 meters.

drainage basin is the catchment area from which a river system obtains its supplies of water. *Precipitation* falls over the area bounded by the major *watershed,* and water makes its way either over the ground surface or by underground routes to the various streams that then converge to form the main river.

dredging is the removal of sand, silt and mud from the bottom of the sea or of a river in order to make it easier for navigation. In the case of river management, dredging is usually a short-term measure as it increases the cross-sectional area, and hence lowers the velocity, which increases the tendency for the river to deposit its load.

drift is the collective name for all the materials (boulders, gravels, sands and clays) deposited under glacial and fluvioglacial conditions. These deposits may be subdivided into:

- *till* – deposited directly by a glacier
- *fluvioglacial* debris – deposited by *meltwater* streams from a glacier.

drought is a lack of rainfall over a long period of time. Droughts occur in many parts of the world but are especially likely in places where the climate is dry and variable, such as desert margins and monsoon areas. One of the worst droughts in recent years has been in the *Sahel* region of Africa. Rainfall has been very low for up to 20 years in some parts, and human activities such as *deforestation* have made the climate even drier. The drought in this area has caused one of the worst *famines* this century. Drought can now be detected from space. Satellite images can show the "greenness" of an area and therefore the level of vegetation cover. A negative change in this vegetation cover can indicate a lower amount of rainfall, which can then alert national governments and relief agencies to possible future food shortages.

drumlins are smooth elongated mounds of unsorted boulders, sands and clays (*till*) deposited by the action of ice. They may be over 50 m in height, 1000 m in length and 500 m wide, with their long axis parallel to the direction of ice movement. The steep, blunted stoss end faces the direction of the ice movement, while the downstream lee end is much more streamlined. Their length is invariably greater than their width, and they usually occur in large numbers or "swarms."

The formation of drumlins is still the subject of debate. One explanation is that they represent a period when a glacier became overloaded with debris, and was forced to shed some as it moved along. This could have taken place at a point where the ice movement slowed due to valley widening. Subsequent streamlining and molding by later movements of cleaner ice would have then taken place. Some drumlins have a rock core that may have impeded the movement of the ice and resulted in deposition. Another theory is that they represent fluctuations in the extent of an ice sheet or glacier. Thus, they are the product of ice readvancing over its own deposits on more than one occasion.

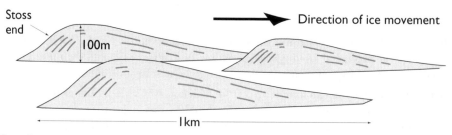

Drumlins

dry adiabatic lapse rate (DALR) refers to the rate at which the temperature of an unsaturated body of air falls as it is forced to rise. The temperature fall is caused by heat loss due to the expansion of the body of air. The DALR remains constant at 9.8°C per 1000 m, or approximately 1°C per 100 m. (See also *adiabatic*, and *saturated adiabatic lapse rate*.)

dry valleys have the typical V-shape and winding pattern of a river valley, but do not have a stream in their bottom. They are common features of chalk and limestone landscapes. However, during wet weather many dry valleys do have seasonal streams called bourns in them, particularly in their lower stretches. There are a number of possible explanations for their existence, of which the main ones are:

- during interglacial and postglacial times the climate was wetter than present-day times, and this caused the *water table* to be higher. Consequently, streams could flow in the valleys that are now dry
- following the end of the last ice age, the ground would have been frozen and therefore *impermeable*. The valleys were thus cut by *meltwater* streams flowing over this frozen ground
- chalk areas were once covered by impermeable rocks upon which rivers flowed and excavated valleys. These were then superimposed on the present landscape.

dual economies: in the *developing* world this refers to a country that contains one or two economically developed areas (*core* and subcores) with regions that surround them being economically poorly developed (*periphery*). Good examples are Brazil and Nigeria.

dumping describes the selling of a good in another country at less than its cost price. A country may dump for a number of reasons:

- to earn foreign exchange – this was a common practice among countries formerly under communist control that were attempting to earn Western currency
- to get rid of excess production – European steelmakers in the late 1980s and early 1990s tried to dump steel in the USA
- to try and destroy foreign industry in order to create a market for the high price goods that are being produced.

dunes are ridges of sand found both in arid areas of the world and along coastlines. Examples of sand dunes in arid areas are the *barchan* and *seif* dunes.

Coastal dunes are formed by the wind on the landward side of a beach. They form best where there is a wide foreshore that dries out between the tides. Strong onshore winds dry out the beach and remove large quantities of sand with which they build dunes. At first embryo dunes develop, which become colonized and stabilized by *marram grass*. These then join up to create larger foredunes that frequently lie parallel to the shoreline. With time, other grasses, fescues and heath plants then

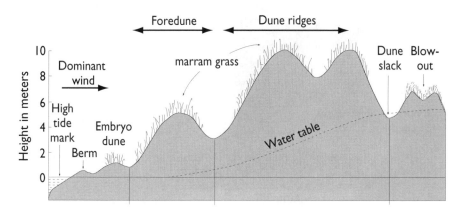

Cross-section across coastal dunes

colonize the dunes, producing a *psammosere*. Exposed areas of sand may become eroded by the wind to create blow-outs.

duricrust refers to the hard crust formed on the surface of the ground resulting from the accumulation and cementation of salts. It is a common feature of arid and semiarid areas where high temperatures bring groundwater rich in dissolved salts to the surface by *capillary action*, the water then being evaporated. Duricrusts can be classified according to the nature of their chemical composition:

- calcretes – rich in calcium carbonate, they are the most widespread and can be several meters thick
- silcretes – rich in silica, and commonly occurring in southern Africa and Australia
- gypcretes – formed of calcium sulphate, most common in very arid areas.

dust bowl: a region of environmental degradation brought about by inefficient farming methods that covered an area across the High Plains, including the Dakotas, Nebraska, Kansas, Oklahoma and the west of Texas. Farmers initially plowed the treeless grasslands mainly for cereals, but a long continuation of this with an overexploitation of the soil led to disaster in the early 1930s when after years of drought, high winds stripped off the topsoil. Many farmers were forced off the land and migrated westwards to California. The story of one Oklahoma family is vividly told in John Steinbeck's novel THE GRAPES OF WRATH. *Soil conservation* methods have now been successfully applied to the region. The term can be used to cover other parts of the world that have been similarly affected.

dykes are vertical intrusions of magma that cut across the *bedding planes* of *sedimentary rocks*. The *magma* cools slowly although those parts that come into contact with

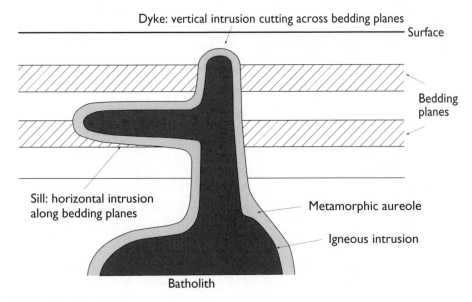

A dyke, sill and batholith

the surrounding rock cool more rapidly to produce a chilled margin. Most dykes are more resistant to erosion than the surrounding rocks and therefore tend to stand out as ridges across an area. Dykes may also refer to ditches or natural water-courses in fenland areas, and also to embankments used to prevent flooding, as in the Netherlands.

dynamic equilibrium refers to the balanced state of a *system* when opposing forces, or inputs and outputs, are equal. If one element in the system changes because of some outside influence, this upsets the equilibrium and affects other components. This is known as *feedback*, which may be *negative feedback* or *positive feedback*.

earthflow: a form of *mass movement* in which debris moves downslope. When weathered material becomes saturated, internal friction between the particles is reduced and stress can cause the debris to move under gravity. This can occur on slopes as gentle as 5° once mobile, but usually needs a slope of about 10° to initiate movement. Flows are generally faster than *creep*.

earthquake: a series of vibrations and shock waves that is initiated by volcanic eruptions or movements along the boundaries of oceanic and continental plates and can occur along *constructive, destructive* and *conservative plate margins*. The point of origin of an earthquake is called the *epicenter* or focus. Shallow focus earthquakes are associated with both constructive and destructive margins but intermediate and deep foci earthquakes only occur where there is *subduction* beneath a destructive margin. There are three types of shock wave: Primary (P) waves, which are the fastest; Secondary (S) waves; and Surface (L) waves, which are the slowest. Seismic recording of the time lag between the arrival of P and S waves enables the identification of an earthquake epicenter. The energy released by an earthquake, the magnitude, is measured on the 10-point *Richter Scale*. The severity of ground movement, the intensity of the earthquake, is measured on the 12-point Mercalli Scale.

EC: see *European Community*.

ecology is the study of the interrelationships between plants and animals and their environment and is concerned with the processes at work in the *ecosystem*.

economically less developed countries (ELDCs) are those countries whose economic growth and therefore development are not as great as the more industrialized countries of the world (*EMDC*). Originally, these countries were known as *developing*, but a general distaste for this term led to its replacement by the *Third World* and, after the *Brandt Report* (1980), the *South*. The nations grouped together under these blanket terms have shown a widening spread of wealth and living standards. There are the *least developed countries (LDCs)*, the poorest countries of the world (mainly to be found in sub-Saharan Africa); the oil-rich states of North Africa and the Middle East; and the *newly industrialized countries (NIC)* particularly to be found in Southeast Asia and Latin America. All these definitions are based on economic growth, and therefore wealth, but there are an increasing numbers of geographers who are looking for other ways of making definitions of such areas based on criteria other than economic.

To group so many nations within one category is also difficult as they vary so much on the following features:

- size of area and population
- historic and colonial background
- physical and human resources
- employment structures
- dependence on external economic and political forces

- political structures
- relative importance of *public* and *private sectors.*

economically more developed countries (EMDCs) are the richest countries in the world with economies that have grown, usually, through industrialization. They were formerly known as the "developed countries" and later as the *First World/Second World* and, after the *Brandt Report,* the *North.* Within this group there is a large difference in economic development, and therefore wealth, between the industrialized countries of North America, Western Europe and Japan (First World) and those countries that were formerly under communist control with *centrally planned economies* – the countries of Eastern Europe and the former *USSR (Second World).*

economic growth describes the way real incomes per head increase over time, a nation's growth being measured in terms of *gross national product.* The reasons for growth are complex, although it can be explained in terms of increasingly efficient uses of the *factors of production* to provide more and more goods and services.

economic man is an assumption that human behavior is based on rational economic motives such as the desire for financial gain. It was one of the assumptions made in many of the *models* in economic geography such as the agricultural land use model proposed by Johann Heinrich *von Thunen.*

economic rent or locational rent is the difference between the revenue obtained by a farmer for a crop grown on a particular piece of land and the total costs of producing that crop and transporting it to market. Economic rent is therefore the profit that can be made from a unit of land. It was one of the concepts put forward by Johann Heinrich *von Thunen* in his model of agricultural land use.

FORMULA $ER/LR = Y(m - c - td)$
where:
ER/LR = economic/locational rent
Y = yield per unit of land
m = market price per unit of commodity
c = production cost per unit of land
t = transport costs per unit of commodity
d = distance from the market

economies of scale are the factors that cause average costs to be lower in large-scale operations than in small-scale ones, therefore doubling the output results in a less than double increase in costs. In such a case, the cost of producing each unit falls because inputs can be utilized more efficiently. Economies of scale fall into two groups, internal and external. Internal economies are:

- specialization – with a large workforce it is possible to divide up the work processes and recruit people whose skills exactly match the job requirements. The workforce is then generally more effective
- fixed costs of equipment – can be spread over more units of production
- purchasing economies with the benefits of bulk buying – obtaining raw materials and components at lower unit costs
- financial economies stem from the lower cost of capital charged to large firms by the providers of finance.

External economies of scale are the advantages of scale that benefit the whole industry, and not just individual firms. If an industry is concentrated in one geographical area then:

- a pool of labor will be attracted to that area and trained to gain the specialized skills useful to the whole industry
- a large grouping of firms will attract a large network of suppliers whose own scale of operations should also yield lower costs.

ecosystem: a community of plants and animals within a physical environment or *habitat.* An ecosystem can exist at any scale as a natural unit from a single tree or a freshwater pond to a *climax vegetation community* or the entire Earth. The various components of the ecosystem such as soils, vegetation, climate and animals are interrelated; energy flows occur in the system and materials such as nutrients are recycled and circulated between the soil, plants and leaf litter and as part of the *food chain.*

ecotone: a zone of transition from one major vegetation community to another, the width of the zone depending upon the rate at which the environmental conditions change. For example, an ecotone exists between the deciduous forest areas and the temperate grasslands. There is usually a greater number of species in the ecotone than in the neighboring communities because the most tolerant species of each community can withstand the changes in environmental conditions. In addition, species that are particularly favored by the conditions in the ecotone also develop.

ecotourism is an environmentally friendly alternative form of tourism. The tourist industry is rapidly expanding at a rate faster than the overall world economy and around 500 million people travel each year for tourism purposes. An increasing number are seeking more exotic destinations with different cultures; international tourist visits to developing countries have more than doubled in the last 20 years. Tourism is being increasingly blamed for massive environmental, cultural and social damage: polluted beaches, degraded coral reefs, displacement of local population, low financial return to host country, abandonment of traditional economic activity. Although other terms such as "responsible tourism," "sustainable tourism" and "low impact tourism" have been coined, ecotourism is commonly used, referring to a niche market for environmentally aware tourists; this has become the fastest expanding sector in the industry. While these approaches may be more beneficial than mass tourism, they do have problems; low impact schemes are likely to become more damaging as they become more successful and it is difficult for such small operations to compete with the cheaper deals of the large tour operators. Ecotourism is an option for the wealthy, therefore attempts to limit the effects of tourism must be aimed at improving mass tourism rather than developing alternatives for minority groups.

ecumene: a term used to denote the most densely populated part of the Earth's surface. It has been estimated that about 60% of the surface is ecumene, the rest being described as nonecumene, the sparsely or intermittently populated or uninhabited area. Delimitation of such areas is difficult because high density areas merge into lower density and within the ecumene there are open areas of parkland, forests and agricultural land.

edaphic: a term that relates to the soil properties and factors that affect plant growth and distribution. These include soil texture, structure, organic content, acidity, soil moisture, nutrients and organisms. Although major vegetation regions largely reflect climatic conditions, within these regions local variations are strongly influenced by edaphic factors.

effective precipitation is that part of the total precipitation that is of use to plants. The effectiveness of precipitation depends on its pattern and the temperatures and rates of *evapotranspiration* present at the time when the precipitation is falling. The type of precipitation is also important as long steady periods of rain allow the moisture to *infiltrate* the soil, whereas short, heavy downpours can lead to a lot of surface *run-off* and are therefore less effective for plants.

EFTA: see *European Free Trade Association.*

ELDC: see *economically less developed countries.*

El Niño: a warm ocean current that occasionally replaces the normal cold Peru Current off the Pacific Coast of South America. Due to a change in the atmospheric circulation with the *ITCZ* nearer to the Equator, there is a reversal of the normal circulation and weather in this area. El Niño (the Christ child) usually begins in December, and warm water extends south in place of the northward flowing Peru Current. This can raise sea temperatures by as much as 10°C, giving rise to heavy rain along the normally arid coast. It is not an annual phenomenon, but occurs at intervals of two to seven years, with its 1997–98 appearance one of great severity, causing meteorological problems around the world.

ELR: see *Environmental Lapse Rate.*

eluviation is a general term for the washing out or removal of any material from a soil horizon. It usually involves the downward movement of clay and other fine particles in suspension.

embargo: an order prohibiting trade with a particular country, perhaps imposed because the country has broken international law or conventions.

EMDC: see *economically more developed countries.*

emergent coast: one that results from a fall in sea level and/or uplift of the land. Features of such a coastline include *raised beaches*, relict cliffs and exposed estuarine mudflats.

emigration is the movement of people away from an area or country to live in another. (See also *migration.*)

employment structure refers to the relative proportions of employment in an area in each of the main sectors of employment. The main sectors are:

- the *primary sector*
- the *secondary sector*
- the *tertiary sector*
- the *quaternary sector.*

There are broad differences in the employment structure of *economically less developed countries (ELDCs)* as compared to that of *economically more developed countries (EMDCs).* ELDCs have a much greater proportion of people employed in the primary sector, and few in the secondary and tertiary sectors. EMDCs have very low proportions in the primary sector, more employed in the secondary sector, and high proportions in the tertiary sector. The quaternary sector has become increasingly important in EMDCs in recent years.

enclosure: the movement away from an *open field system* where land was held in common to a system of small rectangular fields held by private landowners. This trend began in medieval times, as early as the 13th century in some parts of Britain, and slowly spread throughout the country, finally as a result of a series of Acts of Parliament between 1750 and 1820. Enclosure has created the patchwork of fields that exists in that country today, bordered by either hedges or stone walls. Later enclosures also took place in previously unfarmed areas such as woodlands and wastelands.

The impact of enclosure on the people of Britain was significant. Many were displaced from the land they had farmed for generations, and many also lost historic rights to graze their animals and to collect wood.

The effect on the landscape was equally great. Much land changed from being arable to pastoral, particularly sheep farming. There was a dispersal of settlement with new farmsteads being built in the middle of the new farms, many well away from the old villages. New straighter roads with wide grass verges were also established between villages, with the occasional right-angled bend indicating the presence of the strips of the former open field system.

endemic: a disease that is habitually prevalent in a certain area or country, and due to permanent local causes.

endogenetic factors and processes are those that are from within the Earth; for example, in landform development, slope formation is affected by factors such as rock type, chemical composition, structure, degree of permeability. These endogenetic factors interact with *exogenetic* processes to produce particular landforms.

energy budget (the Earth): the balance between the incoming solar radiation (*insolation*) and the outgoing radiation from the planet. Since the Earth's geological record shows that it is neither cooling nor warming to any appreciable extent, the two types of radiation given above must be relatively equal. However, there are variations between the Earth's surface and the atmosphere.

The Earth's surface has a net gain of energy, receiving incoming radiation from the sun and downward from the atmosphere. On the other hand, the atmosphere has a net deficit of energy. To compensate for this difference, heat is transferred from the Earth's surface to the atmosphere by radiation, conduction and by the release of *latent heat.*

There are also variations between the different latitudes on the surface. Low latitudes receive a net surplus of energy, whereas high latitudes (polewards of 40 degrees N and S) have a net deficit. Since the poles are not becoming colder, and the equatorial regions are not getting hotter, heat is transferred between the two by means of air movements (winds), and water movements (*ocean currents*).

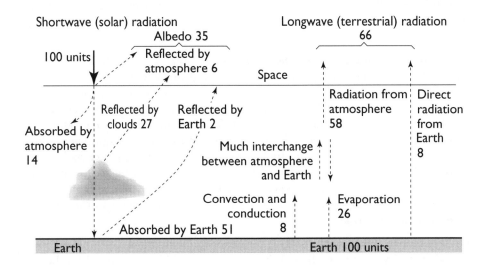

The Earth's energy budget

energy conservation refers to the variety of methods by which the use of all types of energy, but particularly electricity and motor vehicle fuels, is limited or reduced. It may be achieved by:

- greater efficiency – for example, more economic fuel consumption in motor vehicles, cavity wall and roof insulation, low-energy-using light bulbs in the home
- the use of alternative sources of energy that are less wasteful in their methods.

energy sources provide heat and motive power for all form of human activities. There are a wide range of energy sources, and they may be classified as being:

- *nonrenewable* – coal, oil, natural gas, fuelwood and nuclear energy
- *renewable* – *hydro-electric power*, wind power, solar energy, *tidal energy, geothermal energy, biogas, biofuels* and wave power.

englacial: within a glacier.

enterprise culture: a social climate that applauds the profit motive in general and starting a small business in particular.

entrainment: the process by which individual or groups of particles are removed from the bed of a river channel for transportation as either *bedload* or *suspended load*.

entrepreneur: an individual with a flair for business opportunities and risk trading. The term is often used to describe a person with the entrepreneurial spirit to set up a new business.

environment is the natural or physical surroundings where people, plants and animals live. It is very complex, and many factors are involved in its evolution and character, all of which interact with each other. It is also important to recognize that the environment is in a state of constant flux because of the range of forces acting upon it.

environmental impact assessment (EIA) is a process in which the significant effects of a development project are identified, predicted and evaluated. Its purpose is also to ensure that the findings are taken into account during the planning, design and authorization of the scheme. EIA cannot prevent environmental damage, but it informs those who make decisions of the environmentally damaging consequences, so that they may be avoided or reduced. EIA is now a requirement within the European Union for schemes such as major roads, airports, power stations and oil refineries. It may also be necessary for smaller schemes in protected, or politically sensitive, areas. An important aspect of any EIA involves the canvasing of public opinion, the main aim of which is to identify the range of potential problems, and then to identify priorities.

environmental lapse rate (ELR) is the change in temperature with height above a particular place at a given time. It is that which would be measured by a thermometer moving vertically through a still atmosphere. The average value for the ELR is a decrease of 6.5°C per 1000 m. However, this value varies both with height, being lower near ground level, and with time, being higher in the summer season.

environmentalism refers to the growing political concern over a range of issues affecting the world's environment. These include *global warming, ozone depletion, acid rain,* rainforest removal and others. "Green" parties and environmental pressure groups are becoming more influential, at both a national scale (e.g. the Green Party in Germany) and an international scale (e.g. *Greenpeace*).

epeirogenic movement involves relatively gentle raising or lowering of parts of the Earth's crust. These usually operate on a large scale and may be referred to as continent building. Such movements do not produce strong folding of rocks, but some tilting may occur.

ephemeral rivers flow intermittently, or seasonally, after rainstorms and are features of desert landscapes. Despite being short-lived they can generate very high discharges. The torrential rain that falls in such areas cannot easily soak into the ground because the presence of *duricrusts* in deserts creates an impermeable surface that inhibits *infiltration*. The lack of vegetation in such areas also means that the rain reaches the ground without being intercepted.

Some desert lakes (*playas*) are also ephemeral.

epicenter is the point on the Earth's surface directly above the origin or focus of an *earthquake.*

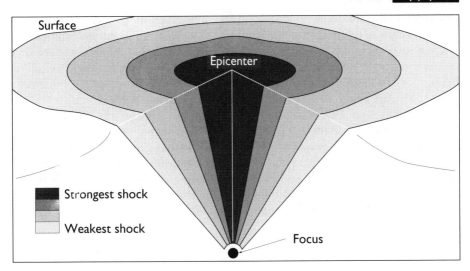

Surface

Epicenter

Strongest shock

Weakest shock

Focus

Epicenter and focus of an earthquake

epidemic: a disease that affects many people simultaneously in a community or area.

epidemiological transition is a model leading on from the simpler *demographic transition model*, which shows how countries go through different stages of diseases and causes of death as they develop. The four stages are:

- age of *epidemic* diseases and famine – mainly infectious diseases, smallpox, malaria, cholera, typhoid
- receding *pandemics* due to vaccines, immunization, improved living conditions
- degenerative and human-induced diseases due to unhealthy diets, stress, smoking, general unfitness (lack of exercise) – cancer, heart disease, respiratory diseases
- degenerative diseases – cancers, Alzheimer's.

The value of such a model is that:

- it sets up a framework within which to set health care strategies
- it can help planners to project future health care needs both nationally and within international health agencies and to decide where best to allocate funds
- it can help manufacturers of medicine and health equipment to project future demand
- it draws attention to the problem of the world's aging population.

epiphytes are plants that develop on the branches and trunks of trees within the tropical rainforest. They are mainly small shrubs and herbs that have escaped the shade conditions on the forest floor by germinating in the tree layer and living on host plants. Many of these epiphytes have a mass of tangled roots that catch falling leaves and debris, which as they decay supply nutrients and act as a sponge to hold and provide a water supply.

equatorial climate occurs in lowland areas within 5° or 10° latitude of the equator and is sometimes referred to as the tropical rainy climate. The annual temperature range is small, normally as low as 3°C, reflecting the fact that there is little seasonal variation in the angle of inclination of the sun. Mean monthly temperatures are of the order of 26°C–28°C. Diurnal range is also small, about 10°–12°C; night temperatures rarely fall below 20°C and day temperatures may rise to 30°–32°C. With its high humidity and monotonous temperatures this climate can be oppressive. Rainfall is high, over 2000 mm per year, and the daily pattern is rather repetitive and predictable. The morning tends to be hazy and as this clears convection currents develop and produce cumulus clouds that intensify during the afternoon heat. Heavy rain occurs in late afternoon or early evening. Convectional uplift is related to the position of the *ITCZ* and therefore a double maxima of rainfall can occur when the sun is overhead at the spring and autumn equinox.

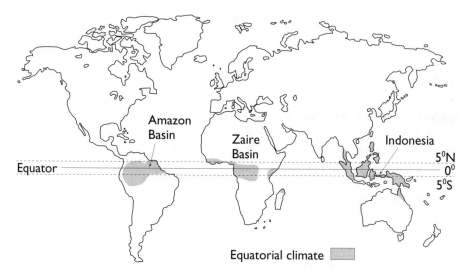

Equatorial climate

erosion involves the removal of weathered material by the action of gravity, water, wind or ice. These agents of erosion transport material that has already been attacked by *weathering*, but they also use the fragments of debris to wear away, or erode, other material through the processes of *abrasion* and *corrasion*.

erratic is the name given to a rock that has been transported by a glacier or an ice sheet and deposited in an area of geology different from that of its source. Therefore it can be used as evidence to indicate the direction of ice movement.

escarpment: the steep slope that forms a more or less continuous line along the margins of an upland area such as a plateau or a cuesta. The term is often abbreviated to scarp. These scarp faces or scarp slopes are the result of differential erosion on horizontal or gently dipping rock strata.

esker: a sinuous ridge of material deposited by meltwater flowing in subglacial channels or through tunnels within the ice. These are called ice contact *fluvioglacial* features. They are formed at right angles to the ice front and consist of silt, sand and gravel deposits that often display some degree of sorting, both from the center of the ridge outward and in the downstream direction, although this may be disrupted by the slumping that occurs as the ice that forms the channel side melts during deglaciation.

estancia: a large farm or estate. It is a term used in the Spanish-speaking areas of South America particularly in relation to the extensive commercial cattle rearing stations.

estuary: the area of a lower river, or mouth, that is affected by tidal change. Estuaries are produced by the postglacial rise of sea level and drowning of former lower valley areas.

ethnic cleansing is a euphemism for the actions of one ethnic group or religious group forcing people of another such group to flee their homes either through eviction or as a consequence of fear and intimidation. It has become a feature of the recent wars in the former Yugoslavia.

ethnic segregation is the clustering together of people with similar ethnic or cultural characteristics into separate residential areas in a town or city. Once a concentration of one ethnic type has become marked it attracts others of a similar ethnicity.

Sadly, such segregation has been beset with problems resulting partly from prejudice, ignorance and intolerance on behalf of the host population. In some *inner city* areas, *deprivation* due to low rates of employment have also helped to create many social problems and tensions, which have surfaced as riots, for example in Los Angeles in 1992 following the acquittal of police officers accused of beating Rodney King.

EU: see *European Union.*

European Community (EC): the name for the European Union before it was renamed from November 1, 1993.

European Free Trade Association (EFTA) is the rival organization to the *European Union.* It was set up in 1959 and included the United Kingdom, Norway, Sweden, Denmark, Austria, Portugal and Switzerland. It was enlarged with the inclusion of Iceland and Finland, but the UK and Denmark left in 1973, Portugal in 1986 and Austria, Finland and Sweden in 1995 when they joined the *European Union.*

European Union (EU): known as the European Community prior to November 1, 1993, and before that as the European Economic Community. It currently consists of the following members: France, Germany, the Netherlands, Belgium, Luxembourg, Italy (the first six members), the United Kingdom, Denmark, Ireland (joined 1973), Greece (1981), Portugal and Spain (1986), Austria, Finland and Sweden (1995). The EU was established under the Treaty of Rome in 1957 with the objective of removing all trade barriers between member states. The original desire was to form a political and economic union that would prevent the possibility of another war in Europe. The EU has a number of institutions:

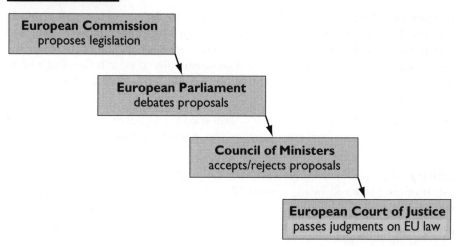

The institutions of the EU

In 1991, agreement was reached with the *EFTA* countries to form closer links within what is now known as the *European Economic Area (EEA)*. This agreement builds on the Single European Act, which came into force in 1987. The *Maastricht* Treaty, which came into force in November 1993 laid the foundation for even greater unity as it was seen as a move toward a federal Europe. Arguments for and against the Union include:

Pros:
- huge potential market of around 350 million people
- the combined strength of the members forms a powerful *trade bloc*
- inward investment is encouraged
- some income and wealth redistribution throughout Europe
- greater freedom for workers within a wider job market

Cons:
- poor distribution of Union income particularly as the *Common Agricultural Policy* takes so much of the budget
- overproduction within agriculture with some corrupt practices
- overbureaucracy within the *European Commission* has brought into question the centralization of law-making
- individual states, in some cases, still put their own interests before the wider community. European laws, therefore, have sometimes been sporadically applied.

eustatic adjustment refers to a worldwide change in sea level. For example, the Ice Ages created a universal fall in sea level since water was stored as ice on the land. Similarly, the melting of the ice sheets at the end of the Ice Ages caused a worldwide rise in sea level as large quantities of water returned to the sea.

eutrophication is the nutrient enrichment of water in rivers and lakes by the accumulation of chemicals from fertilizers and/or slurry from farms and farmland. Farmers use fertilizers to produce healthy crops and to increase yields, but if too much nitrogenous fertilizer is used, or too much animal manure is added to the soil, then some remains unabsorbed and may be leached to contaminate rivers and lakes. Due to the nutrient enrichment in the water, algae and other *phytoplankton* multiply rapidly to produce *algal blooms*, using up oxygen in the water and blocking out the light. Further bacteria multiply to decompose these algae, and use up even more oxygen. Consequently other organisms such as fish are starved of oxygen and die.

evaporation is the process by which liquid water is transformed into water vapor, which is a gas. A large amount of energy is required, usually provided by heat or by the movement of air.

evapotranspiration is the total amount of moisture removed by evaporation and transpiration from a vegetated land surface. (See also *potential evapotranspiration.*)

exfoliation: a process of *mechanical weathering* where the outer layers of the rock peel away like the layers on an onion. Some texts actually refer to this feature as onion skin weathering. The process was thought to be the result of the heating of the outer surface of an exposed rock surface, which would heat faster than the inner areas, leading to a greater expansion and contraction rate. Stresses were therefore set up between the outer and inner parts of the rock, which led to cracking and the outer layers peeling away. A number of laboratory experiments, however, cast doubt on this process, and it would now appear that water has to be present before the rock will behave in this way. The process is therefore probably connected with *chemical weathering*, one likely explanation being that water causes the minerals in the rock to swell (*hydration*) and then the outer layers gradually peel away when rapid heating and cooling occur.

exogenetic: a term covering all those processes that are at work on the surface of the Earth shaping the land. These agencies are generally responsible for the wearing away of parts of the surface and the subsequent *deposition* of the debris that is created. Such processes include *weathering, mass wasting* and *erosion.*

exotic river: a river that maintains its course through an area that has insufficient rainfall to support the channel flow. The river receives the majority of its discharge from outside the immediate area through which it is flowing; for example, where rivers maintain their flow through desert or semidesert regions they are supplied with water from their source regions, which begin in high rainfall, or snow-fed, mountain areas. The River Nile is fed by the White Nile (from Lake Victoria) and the Blue Nile (swollen by heavy rain in the Ethiopian uplands).

expanded town: an urban center that has been selected by planners to have a substantial increase in size in order to avoid too much growth in larger cities and *conurbations.*

extensive agriculture is where a relatively small amount of agricultural produce is obtained from a large area. *Inputs* per unit of land are low. Extensive can apply to both *arable* and *pastoral* agriculture and can be found at a *commercial* as well as a *subsistence* level. Pastoral types that are extensive include *nomadic* pastoralism, beef cattle ranching and hill sheep farming. In arable farming, extensive agriculture occurs in the Amazon Basin (*shifting cultivation*) and in areas where wheat is grown on a large scale, for example the American Great Plains. (See also *intensive agriculture.*)

extractive industry refers to that part of the *primary sector* in which minerals are removed from the ground. It covers the mining and quarrying industries, e.g. coal, oil, natural gas, iron and other mineral ores, salt, limestone, granite.

extrapolation: in forecasting the near future it can be assumed that the recent past will be a good guide. This is known as extrapolating from past to future. When trends have been established, they can be plotted on a graph and extrapolation by eye or by mathematical means can be undertaken.

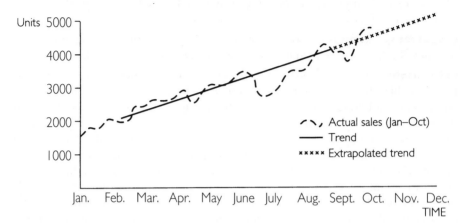

Extrapolation of a sales trend

extrusive: when molten rock is forced to the surface it is known as *lava*, which is said to be extrusive (along with associated material such as *pyroclastics*). This gives rise to a number of extrusive landforms such as lava flows and *volcanoes*. (See also *intrusive*.)

eye (hurricane): an area in the center of a *hurricane* that only develops when the system has reached maturity. This is an area 25–50 km across, with subsiding air producing clear skies, light winds and a temperature higher than the surrounding windy areas.

factors of production include land, labor, capital and enterprise (*entrepreneurship*). These are the factors that combine in order to produce profit.

factory farming is the use of *production line* techniques to ensure maximum output from farm livestock with a minimum of input. This usually means that the animals live in isolation in very confined spaces and are sometimes fattened up with the use of hormones. It has been at times a controversial method of production, with regular protests by animal-rights groups.

famine: it seems rather obvious that a famine is a chronic shortage of food in which many people die from starvation. In recent years though, this traditional view of famine has come to be challenged by many leading authorities. Questions that have been asked include:

- why have some famines occurred when there has been an increase in food availability?
- how is it that famines affect some groups in society but not others?
- why is it that food can sometimes be found in markets in famine-affected areas?
- how is it that food is sometimes exported from famine areas?

A now widely accepted view is that most famines result from a combination of natural events and human mismanagement. Many authorities refer to famine as being the result of a decline in access to food rather than a decline in its availability. People are seen as having endowments of goods and resources and using these to obtain what they want. "Entitlement" failure occurs when an individual's endowment is too small to secure sufficient food for survival. Different groups in society have different endowments; therefore there must be differences in peoples' vulnerability to famine. Definitions here are important, because if famines are simply instances of wide-spread starvation, then food *aid* is the obvious relief. But if they are long-term processes of asset deprivation, then apart from the immediate provision of food aid, the solutions should be directed toward improving the asset situation.

The 1974 famine in Bangladesh was not simply a shortage of food resulting from flooding. The rice crop was not completely destroyed but fears of poor harvests led to large increases in price. Combined with inflationary pressures in the economy, this led to a decline in rural employment and decreased rural wages, leading in turn to a collapse in the entitlement of the rural workforce and their families.

FAO: see *Food and Agriculture Organization.*

fatalism regards hazards as acts of God about which nothing can be done. It is a belief that all such events are predetermined, and the consequences are inevitable.

faulting involves the fracturing of the Earth's crust along which the rocks have been displaced either vertically, horizontally or at some intermediate angle.

The movement takes place along a fault-plane, and the angle between it and the

(a) Types of faulting

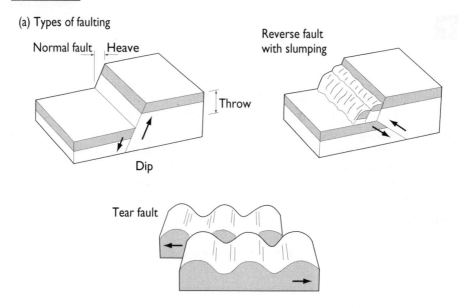

(b) Landforms resulting from faulting

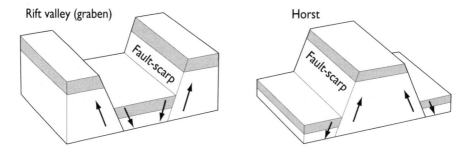

Faulting

horizontal is called the dip. The vertical displacement is known as the throw, whereas the lateral displacement is known as the heave. Different kinds of fault include:

- a normal fault – the result of tension, the rocks being displaced in the direction of the fault plane
- a reverse fault – the product of compression, the rocks of one side of the fault plane being thrust over those on the other side
- a tear fault – where the movement is in a horizontal direction.

federalism occurs when a number of states, which have some *autonomy* or responsibility for governing their own area, combine into a federation for the control of some government functions. National issues such as defense or foreign policy would be dealt with at the national level while decisions on education, housing and health provision may be taken at state level. The USA operates as a union of federal states.

feedback occurs when one element of a system changes because of some outside influence. This upsets the *dynamic equilibrium*, or state of balance, and affects other components in the *system*. (See also *negative feedback* and *positive feedback*.)

fermentation refers to one of the surface layers in a *soil profile* where the decay of freshly deposited plant litter is active. It lies immediately below the litter layer where leaves, twigs and cones have fallen, and the form of some of the plant remains is still visible.

ferralitic soils are found in the humid tropical areas of the world, and are deep and intensively weathered. They result from the high annual temperatures and levels of rainfall that cause rapid *weathering* and growth and subsequent decomposition of luxuriant vegetation. Continuous leaf fall creates a thick litter layer, but the underlying humus layer is thin due to the rapid rates of decomposition by intensive biotic activity. The heavy rainfall causes severe *leaching*. The silica in the soil becomes more mobile than the iron and aluminum *sesquioxides*, and is removed. This leaves a soil rich in iron and aluminum, which gives the soil a deep red color. The continual leaching and abundance of mixing fauna inhibits the development of soil horizons.

These soils are very sensitive to environmental change. Once the supply of nutrients is removed, by *deforestation*, the soils soon lose their fertility. They have a loose structure and, when exposed, are easily eroded by heavy rainfall.

Ferrels law states that a body moving over the surface of the Earth will be deflected to its right in the northern hemisphere and to its left in the southern hemisphere. This is due to the Earth's rotation and the *Coriolis Force*, and is particularly related to atmospheric circulation.

ferruginous soils are found in the tropical regions of the world that experience marked wet and dry seasons (regions known as the *savanna*). As the grasses of these areas die back during the dry season they provide organic matter, which is decomposed to give a thin layer of humus. With high rates of evaporation, *capillary action* also brings bases and salts in solution to the surface. During the wet season, *leaching* removes the silica from the upper horizons, leaving behind iron and aluminum *sesquioxides*, which give the soil a red color. This alternating pattern often produces a hard cemented layer called *laterite* just below the surface of the soil, which impedes drainage, root penetration and plowing.

These soils are not particularly suitable for cultivation, and often form grazing lands. They contain few nutrients, and are vulnerable to erosion by both the heavy rains of the wet season and the action of wind during the dry season.

fertility (population) is the average number of children each woman in a population will bear. It is usual to refer to women between the ages of 15 and 50 in such calculations. If fertility is 2.1 children, then it is likely that a population will replace itself.

fertility (soil) is a statement of how suitable a soil is for the growth of crops. If a soil is fertile, plant growth is rapid and yields are high. Such a soil provides water, air and nutrients in the correct amounts and proportions. A fertile soil therefore has all or some of the following characteristics:

- a neutral to slightly acid *pH* value
- a loamy texture giving good drainage and aeration
- a crumb structure associated with adequate amounts of humus
- a good supply of nutrients either from the decay of organic matter or the addition of chemical *fertilizers*.

fertilizer is used by farmers to produce healthy crops and to increase yields. It is added to a soil in order to replace the nutrients that have been removed by the growth of crops and are not replaced when they are harvested. Fertilizer may be:

- organic – farmyard manure, which is then plowed into the soil. This not only returns nutrients to the soil, it also improves the structure of the soil
- inorganic – chemicals, especially compounds of nitrogen, potassium and phosphorous.

fetch is the length of open sea over which a wind blows to generate waves. The longer the fetch the greater the potential is for large waves. It is possible that some waves may have originated several thousand kilometers away.

finite is a term applied to resources when they are *nonrenewable*. Once they have been used up they cannot be replaced. Minerals and fossil fuels are examples of this type of resource, which is being consumed at an increasing rate as more countries progress toward industrial development.

fire hazard occurs when there is some threat to the natural, built or human environment as a result of fire, whether due to natural or human causes. The term tends to relate to natural fires rather than those that are the result of domestic or industrial accidents, and as such can be regarded as an ecological hazard. The brushwood and coniferous vegetation of the *Mediterranean* regions of France, combined with the long, hot dry summers, provide ideal conditions for fires, which once ignited are often fanned southwards by the mistral, a dry wind that blows down the Rhone Valley toward the coast. This threatens the vegetation, the urban areas, the people and the economy of the Riviera coast.

First World: the economically more advanced countries that have some form of free market economy. This includes countries in North America and Western Europe as well as Australia, New Zealand, Japan and Israel.

fish farming involves the deliberate feeding, breeding and harvesting of fish in man-made pools or tanks but also includes the use of enclosures in river channels. Careful monitoring is needed to avoid losses from disease in such confined environments; regular application of chemicals and antibiotics has enabled effective control.

fissure: a line of weakness, such as a joint or a fault, in crustal rocks through which volcanic lava may erupt. Such action usually involves more basic, fluid lava accompanied by little explosive activity.

fjord: a long, narrow, steep-sided coastal feature formed by the drowning of a glaciated valley. They usually result from valley glaciers eroding below the level of the sea, producing an overdeepened U-shaped valley. This valley has been inundated by the postglacial *eustatic* rise of sea level and in long profile often reveals a series of glacial basins that become shallower toward the seaward end of the fjord where a submerged threshold is found. This can occur as a major rock bar, sometimes with moraine, and it marks the point where the lower section of the glacier reached the sea and began to float, thus reducing the vertical erosive power of the ice. Evidence for *isostatic* uplift in postglacial times may be seen in *raised beaches* at the head of some fjords. These coastal features are well developed in Norway, but are also found in British Columbia, South Island, New Zealand and southern Chile.

flash flood: a large but temporary increase in channel discharge. The *hydrograph* is modified and is characterized by a steep rising limb, a short lag time, a brief period of peak flow and a gentle recession limb. These short-lived floods are particularly associated with deserts but they can occur in other environments when heavy rainfall follows an extended period of wet weather that has saturated the soil. With reduced *infiltration*, faster *overland flow* transports water into channels more quickly, giving a rapid increase in discharge.

flexible manufacturing systems make a wider range of specialized products than traditional (*Fordist*) industries. They are able to adjust to alterations in market demand; to achieve this they rely on many different suppliers to provide materials and components as they are required. These firms also use computerized machinery that can be adapted to make a variety of products, unlike traditional firms where there is mass production for a mass market. Labor is also more flexible and less unionized, and, unlike the demarcation and division of traditional factory working, a smaller workforce operates multitasking in which one person is capable of undertaking a number of jobs as production targets demand.

flocculation is the process by which a river's load carried in suspension is deposited more easily upon its meeting with sodium chloride in sea water. The meeting of fresh and salt water in river estuaries and deltas causes a clustering or coagulation effect on silt and clay particles, and these larger particles sink more rapidly.

flood control seeks to reduce the frequency and magnitude of *flooding* that takes place and therefore limit the damage that floods cause. There are two major approaches to flood control:

1 flood protection. Achieved by:
   - modifications to the banks and/or channel to enable the river channel to carry a larger volume of water. Artificially raised and strengthened banks form a significant part of this strategy. In some cases parallel lines of such floodbanks act as a double form of protection – if the river overtops the first barrier, then it has difficulty rising over the second bank some distance behind. The removal of large boulders from the bed of the river reduces roughness, therefore increasing the velocity of flow
   - the building of dams to regulate the rate at which water passes down a river
   - diverting rivers away from vulnerable areas
   - increasing the height of the floodplain by dumping material on it.

2 flood abatement. Achieved by:
   - *afforestation*, which slows down the rate at which water reaches a river, as well as reducing the amount that actually does reach it

- contour plowing and strip farming in semiarid areas that reduce the amount of surface run-off, and therefore reduce the liability to flooding.

flooding occurs when a river's *discharge* exceeds the capacity of its channel to carry that discharge. The river overflows its banks. Flooding may be caused by a number of factors:

- excessive levels of precipitation
- the melting of snow
- the failure of man-made dams and/or embankments
- the changing of land use in catchment areas such as *deforestation* and urbanization.

Flooding may also take place in coastal areas due to either rising sea levels or tidal surges caused by storms.

Bangladesh, being located at the mouth of the Rivers Ganges and Jumana, and at the head of the Bay of Bengal, which suffers from tropical cyclones, is often trapped between two sets of floods – one caused by rivers, and the other by the sea.

flood plain: part of a valley floor that a river may flood from time to time. As the river floods it deposits a layer of silt that gradually builds up the height of the flood plain. The edge of the plain is often marked by a prominent slope known as a *bluff* line. Flood plains may be found throughout a river valley, even extending in restricted fashion into upland areas.

flora and fauna: flora is the name given to the plant species that make up the vegetation of an area, while fauna is the name given to the animal species.

flow line: a technique for presenting data in which the width of the line between two points in the network is proportional to the volume of movement. The route followed by the line represents the line of movement of the people, traffic, goods, information or other data being presented, whereas a *desire line* simply shows the origin and destination of the movement.

flow production is the manufacture of an item in a continuously moving process. Each stage is linked with the next by a conveyor belt or in liquid form, so that the production time is minimized and production efficiency is maximized. The

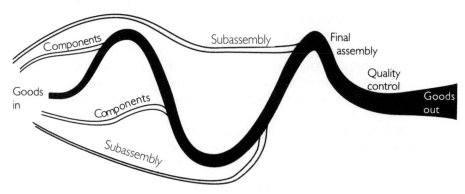

Continuous flow production

accompanying diagram shows a continuous system in which subcomponents are being fed into the main production line just as streams flow into a river.

fluvial is a term applied to the action of rivers. Fluvial erosion includes mechanical *abrasion*, or *corrasion*, when fragments of debris are rolled and dragged along the bed of the channel, slowly wearing away other material. The sheer power of water flow may lead to *hydraulic action* eroding material. Fluvial transport involves traction, when sediment moves along the channel bed; *saltation*, when pressure differences across particles cause them to be projected temporarily into the main velocity flow that moves them downstream; and suspension, when small particles are carried along in the water flow. As stream velocity decreases fluvial deposition occurs; particles settle according to their grain size with the largest diameter material being deposited first. The relationship between stream velocity and particle size in relation to fluvial erosion, transport and deposition is displayed in the *Hjulstrom curve*.

fluvioglacial (or glaciofluvial) is a term used to describe processes and landforms resulting from the action of meltwater streams associated with glacial environments.

fluvioglacial landforms are produced by the action of meltwater streams transporting and depositing material in a glacial environment. These streams may flow on the ice (*supraglacial*), within the ice (*englacial*) or beneath the ice (*subglacial*) and can carry material beyond the snout of the glacier or the edge of the ice sheet. The resulting landforms can be divided into two groups: proglacial and ice-contact features. Proglacial landforms include *outwash* plains and valley trains that are deposited beyond the ice margin. Ice-contact landforms are produced under or on the margins of the ice and include *eskers*, *kames* and kame terraces. Fluvioglacial landforms consist of stratified deposits, which distinguishes them from the unstratified materials produced as a direct result of glacial deposition.

fluvioglacial processes include the transport and deposition of material already eroded by glacial action. When meltwater streams are flowing within or under the ice they are subjected to greater hydrostatic pressure, which increases their capacity to transport sediments. As the discharge of streams varies seasonally, some material is deposited along the bed and banks of these stream channels. When streams emerge at the ice margin, the decrease in pressure, and resulting fall in velocity, leads to a reduction in the stream's capacity to carry material, and deposition occurs. This results in graded deposits with larger particles nearer to the ice front and progressively smaller particles at greater distance.

fog is a term applied to an atmospheric condition when visibility is less than 1 km. It is caused by the cooling of the air to *dew point* and the resulting condensation of water vapor in the atmosphere at ground level. This cooling can result from three different modes of formation, giving three types of fog: radiation, advection and frontal.

fohn: a warm, dry wind that descends from the Alps. Temperature rises of between 15°C and 20°C may be experienced and this can cause rapid snow melt and avalanche problems. The Fohn effect, which is similar to the changes caused by the Chinook wind in the Canadian Rockies, is caused by low pressure systems to the north of the Alps drawing moist air from the Mediterranean. As this air rises over the Alps it is cooled at the *dry adiabatic lapse rate* until it reaches its *dew point*. At this temperature, when *relative humidity* is 100%, the air becomes saturated and on rising higher it will cool at the *saturated adiabatic lapse rate*. Condensation and precipitation occurs on the windward side of the mountain barrier and as the air begins to descend on the leeward side, it warms up at the dry rate. Because of the difference between the dry and the saturated rates, there is a net increase in temperature as the air crosses the mountains.

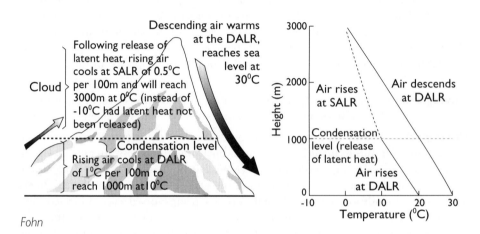

Fohn

folding is the bending and crumpling of sedimentary layers and is caused by compressive forces in the Earth's crust. When rigid blocks move toward each other the sediments in between are crumpled and may be contorted into folds of varying intensity and complexity. These can range from simple monoclines, anticlines and synclines, to more complex recumbent folds, and to extreme forms such as nappes. The forces involved are so powerful that strata may be metamorphosed, leading to the transformation of rocks. For example, clays may be changed into slates. Fracturing associated with the folding may lead to intrusive and extrusive volcanic activity. Fold mountain development is related to the movement of crustal plates at *destructive plate margins*, when ocean crust subducts beneath continental crust or when two continental plates collide.

Food and Agriculture Organization (FAO) was the first new permanent specialized agency of the United Nations, established at the end of World War II with the objective of eliminating hunger and improving nutrition. The FAO seeks to coordinate the efforts of governments and technical agencies in programs for developing agriculture, forestry and fisheries. The headquarters of the organization are in Rome. The work of the FAO includes:

- research into new methods, crops (particularly *HYVs*), pest control, etc.
- providing technical assistance on various projects
- running programs to promote more rural employment
- promoting agricultural exports
- running educational programs

- maintaining statistics on world production, trade and the consumption of agricultural commodities
- publishing a number of periodicals, yearbooks and research bulletins.

food chain: the flow of energy through an *ecosystem*. Each link in the chain feeds on and obtains energy from the one preceding it and in turn is consumed by and provides energy for the following link. Each link in the chain is known as a *trophic level*. There are usually several trophic levels in a chain. At its simplest, this could consist of plants, then herbivores, followed by carnivores and finally omnivores (that feed on both animals and plants and include humans in the group). At each stage in the chain material is lost to decomposer organisms, mainly *bacteria* and fungi, which break down this matter into its constituent parts, making it available for reuse by plants.

food deficiency exists when people consume on average less than 10.8 megajoules per capita per day. The United Nations regards this amount as being the minimum amount of food needed for healthy living. The *Food and Agriculture Organization (FAO)* states that there is enough food grown in the world to feed everyone, so there should not be a food deficiency anywhere. The problem is one of uneven distribution and the increasing cost of buying and transporting food.

People who do not eat enough food are said to be undernourished, whereas those who do not eat the right diet suffer from malnutrition. Both of these conditions may lead to a variety of *diet* deficiency diseases.

food marketing channel refers to the institutions and routes involved in the production, distribution and consumption of food. The "traditional" marketing channel for food has been from the farmer to a wholesaler, and then to a retailer. However, in recent years there have been a number of major changes to these channels:

- the growth of *contract farming*
- producer cooperatives
- direct marketing through farm shops, roadside stalls and Pick Your Own.

food processing involves the preservation of food so that it can be sold either further away from the point of production, or longer after its production. The main forms of food processing are freezing, refrigeration and canning. The creation of prepackaged convenience foods forms an increasing proportion of food processing activities.

Benefits resulting from food processing for both the farmer and the consumer are:

- farmers receive a greater degree of income security
- consumers receive regular supplies of food, as well as a greater variety of foods.

food retailers are the organizations through which food is sold. They exist at varying scales from the small local store to the larger supermarket. Much food used to go from farms to wholesale markets and then on to the smaller retail outlet. Nowadays, a larger proportion goes directly to the supermarket chain as a result of *contract farming*. Farmers are also retailing their produce themselves with the development of farm shops and Pick Your Own.

food surpluses occur when production has exceeded demand. They have been a common feature of the European Union *Common Agricultural Policy (CAP)* due to its system of guaranteed prices, which has encouraged farmers to overproduce. "Mountains" of food have been created in grain, beef, butter and yellow raisins, in addition to "lakes" of milk, wine and olive oil. Many of these surpluses are either placed in storage at great expense, or sold at cheaper prices to other countries, such as those of Eastern Europe.

food web: a complex series of food chains where organisms operate at more than one feeding or *trophic level*. Any animal's diet is usually quite varied, and one species of animal or plant may be part of the food of a wide range of different animals. Consequently, complex interactions of feeding are set up. Humans operate at several trophic levels by being both herbivores and carnivores, and therefore complicate any simple *food chain*.

footloose industries are those that have a relatively free choice of location. They are not tied to the location of raw materials. Many are the newer industries that either provide a service for people or produce light goods of a high value, and are therefore market-oriented. Examples include high-tech and electronics industries, food processing and distributive companies. Footloose industries are frequently located in large new industrial parks on the edges of towns and cities, or alongside major highways to utilize the efficient road transportation system.

Fordism: the application of Henry Ford's faith in mass production, run by an auto-cratic management. This involved the division of labor, i.e. the breaking down of a job into small, repetitive fragments, each of which could be done at speed by workers with little formal training. The benefits for production were outstanding and costs dramatically declined. By the 1960s, however, such methods were coming under increased scrutiny as people demanded more satisfaction from their employment.

foreset beds are layers of sediment that are laid down at the seaward edge of a *delta*. They consist of mainly clays and silts, which are deposited at an angle as the delta builds outward. The coarser sands and silts are deposited first in the *topset beds*, and even finer clays are carried furthest to form the *bottomset beds*.

forest clearance: the removal of woodland in order to establish land that could be used for arable production, pasture or village settlement.

forest management is the deliberate preparation, planting and cutting of forest areas. It is a planned process of felling and replanting; as one section of the forest is cut new areas are plowed over and planted. The land is prepared by ditching to lower the water table, and fertilizers are used to encourage growth. The trees are protected from pests and disease and firebreaks are left between sections of the forest. Dead or diseased trees are removed and sections are thinned out to provide optimum conditions so that fully mature trees can be felled.

Fortress Europe describes the possibility that a Europe with a single market might build an import protection wall around itself, keeping out American and Japanese imports.

forward integration: a type of vertical integration in which a company gains con-trol of activities downstream of it, i.e. those firms that deal with the finished product of the firm in question. For example, car manufacturers could seek to gain control of car distributors and retailers. (See *vertical integration.*)

fossil fuels are those fuels, consisting of hydrocarbons, that were laid down in past geological periods. They include coal, oil and natural gas, and are classified as *non-renewable resources*. The bulk of the world's energy is produced from them, and this reliance is likely to last well into the 21st century.

Pros:
- they are widespread in their occurrence and therefore accessible to many nations
- they have been converted economically for a long time into forms of energy
- they are readily converted from one form to another, i.e. from liquid to gas, from solid to gas
- they make excellent fuels for transport
- there are several hundred years of recoverable coal deposits

Cons:
- they are nonrenewable
- oil and gas reserves are far more limited than coal
- they are also a valuable source of raw materials for a wide range of chemical products and many feel that they should not be "squandered" by being used as fuels
- burning fossil fuels contributes to atmospheric pollution such as *smog*
- they are a major contributor to *acid rain*
- the release of *carbon dioxide* is a major contributor to the *greenhouse effect* and thus *global warming*.

fragmentation is when agricultural land is broken up so that one farm may consist of numerous small and scattered fields. This is usually the result of inheritance laws when the land is divided up between the sons or land is given in the dowry of a bride. Fragmentation wastes a lot of the farmer's time as he moves between fields and does not promote an efficient farming system. *Mechanization*, for example, is seriously hindered.

freeport: an area set aside for commercial activities where costs are saved as the area is not considered to be within the country for taxation purposes. Therefore raw materials can be imported and made into finished products, which are then exported, all without attracting any taxation or duty of any kind. Singapore, for example, has seven such zones, six based around its docks and the seventh at its airport. Goods can be manufactured or simply assembled, with no import or export duties being paid. Profits can be sent to the parent country of the companies involved, again free of taxation. The main advantage to Singapore is the employment that is created by these zones, together with the economic stimulus this gives to Singapore's economy through the increased spending power of those employed within the zones.

free trade exists when trade between countries is not restricted in any way by *tariffs*, *quotas*, or other barriers. It is based on the theory that every country will be better off if it specializes in producing goods at which it is comparatively more efficient. Until its dissolution at the end of 1995, the *General Agreement on Tariffs and Trade (GATT)* accepted the notion that more trade benefits everyone, and through successive "rounds" of negotiations sought to reduce trade barriers around the world.

freeze-thaw action: see *frost shattering*.

frequency distribution: see *dispersion measure*.

friable: a description of a soil that is easily broken up.

frictional unemployment occurs in the time delay between losing one job and finding another. By its nature it is temporary, as opposed to *structural unemployment*, which is more fundamental and therefore more long-term.

friction of distance refers to the lesser likelihood of people using a service, the greater the distance away that they live from it. Distance is perceived to be a disadvantage, due to either the time, cost or effort involved. The effect of the friction of distance is to create a *distance decay* in the use of a service.

Friedmann, John produced a model of the economic development of a country, with particular reference to the changing spatial relationships within that country. The model progresses through four stages of development for a country:

Stage 1

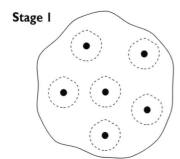

Relatively independent centers, no hierarchy.
Each town lies at center of a small region.

Stage 2

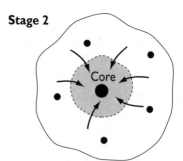

A single strong core, with a periphery.
Labor and capital move to the core.

Stage 3

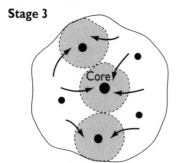

A single national core, with a number of peripheral subcores.

Stage 4

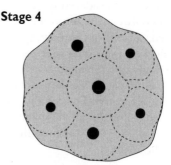

An interdependent system of cities, with a maximum growth potential for the whole country.

Friedmann's development model

- Stage 1 – a number of relatively independent local centers exist, each of which serves a small region, with no settlement hierarchy
- Stage 2 – the development of a single strong core during the initial phases of industrialization, with an underdeveloped *periphery* in the remainder of the country. Development occurs in the core region, which has a specific advantage over the rest of the country, for example a natural resource or dense population. The initial advantage is maintained by *cumulative causation* as more capital, entrepreneurs and labor move to the core
- Stage 3 – the core-periphery structure becomes transformed into a multi-nuclear structure with the national core and a number of peripheral sub-cores. These may develop due to large regional markets or important natural resources
- Stage 4 – a functional interdependent system of cities resulting in national integration and maximum growth potential.

From this model four types of areas can be designated:

- the core region – the focus of the national market, and seedbed of new industry and *innovations*
- upward transitional areas – regions with some form of natural endowment characterized by inward migration of people and investment
- downward transitional areas – regions with unfavorable locations and resource bases, characterized by outward migration of people and investment
- resource frontiers – areas where new resources are discovered and exploited.

fringing reef: an area of *coral* that is attached to the coast and extends out to sea. It represents an early stage in the sequence of subsidence or sea level rise, which leads to the formation of *atolls*. As relative sea level rises the coral continues to grow upward and away from the subsiding land, producing first a barrier reef and then an atoll.

front: a boundary separating two *air masses* with different temperatures and densities. The term dates from World War I when meteorologists compared such a meeting of contrasting air masses to the trench warfare of that time. There are two main types of front:

- a *warm front* – when warmer air is advancing and rising up and over a wedge of cold air ahead of it. Although the gradient of the incline is gradual (1°), the rising air cools and the water vapor within it condenses to produce cloud and precipitation.
- a *cold front.*

Fronts and their passage over an area are significant features of a *depression.*

frontogenesis: the activity that takes place along a *front* and results in the growth of waves along the front into fully fledged *depressions*. Contact on the front between air masses of different temperatures, humidity, density, speed and direction of movement creates friction, which in turn sets up waves. *Convergence* and uplift at the apex of the wave allows surface pressure to fall, creating the center of the depression. *Divergence* aloft allows the rising air to be moved on (by *jet* *stream*), which is fundamental in deepening the depression and bringing maturity to the system.

frost is the deposit of fine ice crystals onto a surface of grass, plant leaves and walls. It forms under clear, calm, anticyclonic conditions in the winter when there has been a rapid heat loss at night. Water vapor condenses directly by *sublimation* onto these surfaces.

Glazed frost is formed by raindrops falling through a layer near the surface with temperatures below freezing, and onto a very cold surface, producing a solid sheet of ice.

frost hollow: a natural depression or valley where *frosts* occur more regularly than in surrounding areas. They frequently occur as a consequence of a *temperature inversion* in winter associated with anticyclonic conditions. They may also occur due to the influence of *mountain winds* (*katabatic* winds).

frost shattering: a form of *mechanical weathering* in rocks that contain crevices and *joints*, and where temperatures fluctuate around 0°C. Water enters the joints, and during cold nights freezes. As water occupies around 9% more volume than water, it exerts pressure within the joint. This alternating freeze-thaw process slowly widens the joints, eventually causing bits to break off from the main body of rock. It leads to the formation of such features as *scree* slopes and *blockfield*.

fuelwood: the use of wood as a fuel for cooking and heating for many families in economically less developed countries. In many parts of the world, the collection of fuelwood is a time-consuming occupation for the women and children, with them having to travel many kilometers to find sufficient amounts. As the demand for fuelwood increases, more trees are cut down, with damaging environmental effects such as *soil erosion*.

full employment is the level of employment that provides jobs for all those who wish to work apart from those *frictionally unemployed*.

function is the main reason why a settlement was built or continues to exist. Different settlements have differing functions, for example:

- some are market towns, where people can buy and sell produce
- some are ports, where goods are either imported or exported
- some are industrial, where the main form of employment is in manufacturing industry
- some are resorts, where tourism is a significant factor.

Settlements usually have a combination of functions, and their functions may change over time. Consequently, the type of classification given above is seen by many as being too simplistic.

functional hierarchy: a method of classifying settlements by their main activity into a tiered structure of importance. An example of a functional hierarchy is:

- (a) rural settlements – market towns, dormitory settlements
- (b) older 19th-century settlements – mining towns, manufacturing towns
- (c) 20th-century settlements – administrative towns, commercial towns, service towns, resort towns, new towns.

Such classifications are based on the determination of a town's function, but this is difficult to define, and judgments are subjective. Also, most towns and cities in developed countries are multifunctional, with no single function predominant. It must also be noted that some settlements grew for a reason that no longer exists, and also that functions change over time.

functional interdependence represents the final and ideal stage of John *Friedmann's* development model. All of the cities in a country are fully integrated into a complex system of economic production and communication. The country should be able to adapt to any change in circumstances, as it has the most efficient locational pattern of cities, and has the maximum potential for further growth.

functional zone: district of a town or city where one type of land use is dominant. The principal functional zones of a town or city are:

- the *Central Business District (CBD)*
- industrial areas
- residential areas.

Each of these zones can be further subdivided into subzones. For example, residential areas could be subdivided into areas of inner city housing, high-rise apartment buildings, single-family houses, and post-1960s suburban communities.

fungicide: toxic chemicals used in *arable farming* in order to eradicate fungal diseases that attack plants.

G7: see *Group of Seven.*

G8: see *Group of Seven.*

Gaia: the theory that the functioning of the planetary *ecosystem* is determined for its own long-term good by the sum total of living and nonliving components of the system. The theory was first put forward by James Lovelock, who took the name Gaia from the Greeks' name for the Goddess of Earth. Lovelock maintained that the planet is not an inert cinder beyond the influence of its organic passengers. Just as it determines their fortunes, they serve to shape its make-up. Lovelock did not set out to show that all this was done willingly, but maintained that organisms work in accord with their surroundings, so that they do not have to know what they are about for them to work beneficially with each other. Ideas on Gaia are beginning to gain some measure of acceptance, particularly because of the way in which man is increasingly responsible for polluting the atmosphere and the oceans, deforesting vast areas and expanding the deserts.

garrigue: the vegetation of the *Mediterranean climatic regions* that occurs on drier and more *permeable* rocks such as limestone. It is lower and less dense than the more typical Mediterranean vegetation of *maquis.* The more common plants are gorse, and a range of aromatic shrubs such as thyme, lavender and rosemary. It is a good example of a *plagioclimax* as much of the vegetation is the result of people's interference.

GATT: see *General Agreement on Tariffs and Trade.*

gavelkind is a system of land inheritance in which, on the death of the landowner, the property is divided equally between the sons. The land may be passed on to daughters if there are no surviving male heirs. Over a few generations this system leads to *fragmentation* of land holdings, which can produce plots too small to be economically viable. It contrasts with *primogeniture* in which land is passed to the eldest son.

GDP: see *gross domestic product.*

gender: the socially or culturally defined difference between men and women, while sex is the biologically defined difference. Gender differences are based on socially defined roles of men and women and the relations between them, such as the degree of power they exert over one another. They have different degrees of access, for example, to certain jobs, higher education and some social and recreational organizations. Some modern observers have noted that many women are constrained in their activities and are limited to the "private domain" of domestic life, while men dominate the "public domain" of paid work and public decision-making.

General Agreement on Tariffs and Trade (GATT) was established after World War II to encourage the growth of international trade by removing or reducing *tariff* and nontariff barriers. Agreements were reached after what were known as "rounds" of negotiations, which often lasted many years. The seventh and last was the Uruguay round. GATT was dissolved at the end of 1995 and absorbed into the newly formed World Trade Organization.

general atmospheric circulation is the pattern made within the atmosphere of winds and *pressure* belts. Although the circulation is extremely complex, there are certain movements that occur regularly enough for us to recognize patterns of air pressure distribution and winds.

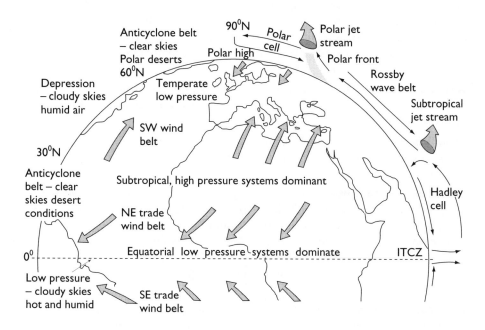

Simplified atmospheric circulation in northern hemisphere

Our modern understanding began with the work of Edmund Halley (1686) followed by George Hadley in 1735, who recognized that a global *convection* system was set in motion by a tropical heat source. That the system was fundamentally a three-celled one was discovered by William Ferrel in 1856 and in more modern times refined by Carl-Gustaf Rossby in 1941.

gentrification: a process of housing improvement associated with a change in neighborhood composition when lower income groups are displaced by more affluent people, usually in professional or managerial occupations. This is one of the processes that can regenerate the *inner cities*. It essentially involves the rehabilitation of old houses and streets, often in areas originally developed in the early part of the 19th century. Gentrification often depends on the cooperation of property developers, real estate agents, building societies and local authorities.

Pros:
- contributes to the social balance of inner city areas
- increases residents' purchasing power and therefore the general prosperity of the area
- creates employment in inner cities

Cons: • lower income people are driven into poorer quality accommodation as the privately rented sector shrinks, therefore increasing urban deprivation
 • decline in the social quality of neighborhood life for the original residents
 • friction often occurs between the gentrifiers and the original residents.

Geographic Information System (GIS): a structured framework for acquisition, storage, retrieval and analysis of data within a common geographically referenced spatial system. Such systems are made up of data on various topics (e.g. soil, vegetation) and from a range of sources (e.g. maps) processed by computer. For example, information obtained from *satellites* can be placed in a GIS where maps are produced of an area and layers of different information may be combined so that relationships become apparent. GIS databases are frequently used in environmental and resource management.

geological survey: an investigation into the nature and structure of rocks at a particular location. It may also seek to identify evidence of ancient life forms in the form of fossils.

geophysics: the branches of physics that are concerned with the Earth and its atmosphere. Meteorology, geomagnetism and seismology are all geophysical subjects.

geostationary satellites maintain a fixed position over a point on the Earth's surface, and therefore monitor only that part of the surface. The meteorological satellite METEOSAT 3, launched by the European Space Agency in 1988, is in geostationary orbit over the equator at 0 degrees longitude. It observes the Atlantic, Europe and Africa, producing an image every 30 minutes. Other such satellites are fixed at longitudinal intervals of 70° around the equator, and together monitor the whole globe.

geostrophic winds blow parallel to the isobars in the upper atmosphere. They are the result of the balance between the *pressure gradient* force and the *Coriolis Force* that exist in air movement at such high altitudes. They are unaffected by friction from the Earth's surface. Their existence gives rise to the maxim: "If, in the northern hemisphere, you stand with your back to the wind, low pressure is to your left, and high pressure is to your right."

geosyncline: the name given to a linear depression in the sea floor in which vast amounts of sediments would have accumulated prior to being pushed together and upward to form fold mountains. Such a depression would have had to subside at the same rate at which it was being infilled by sediments. A geosyncline, called the Sea of Tethys, is said to have existed in Eurasia. The sediments in it were pushed upward to create the Himalayan mountain range when the Indian Plate moved north and collided with the Eurasian Plate.

geothermal energy is derived from the heated rocks beneath the surface of the Earth. Cold water is pumped down bore holes, heated by contact with the hot underlying rocks, and turned to steam. It is then returned to the surface, where it is used to generate electricity. Geothermal power is used in a variety of locations around the world such as Iceland, New Zealand and El Salvador. In Iceland, it has enabled heat to be provided to homes, open-air swimming pools and greenhouses. Due to such heat, Iceland is the only European country that is self-sufficient in bananas!

ghetto: originally referred to a concentration of Jews within a city. However, particularly in the United States, the term is now used to refer to a highly segregated area where a minority, often ethnic, group is dominant. The most common forms of ghettos in American cities are those dominated by Afro-Caribbeans and Hispanics. Such areas are characterized by low incomes, high unemployment, substandard housing and high levels of delinquency and crime.

gini coefficient is a statistical measure of the degree of similarity between two sets of percentage data. It is calculated using:

FORMULA $\quad$ Gini $(G) = \frac{1}{2}\Sigma(X_i - Y_i)$

where X_i and Y_i represent the two sets of percentage data. The coefficient ranges from 0 to 100; a value of 0 indicates that the two sets of percentages are identical, i.e. $X_i - Y_i = 0$. A value of 100 indicates that the data is as different as it is possible to be. Because the gini coefficient has a fixed range of values it provides a simple and consistent framework for interpreting results.

For example, differences in the variation of manufacturing employment in a region could be compared with the pattern of manufacturing employment on the national scale.

Manufacturing employment for four industries in two regions and national pattern.

Industry	Region A %		Region B %		National % (Y_i)
W	$(X_i)_A$	20	$(X_i)_B$	10	25
X		10		20	10
Y		40		10	25
Z		30		60	40

For Region A,

$G = \frac{1}{2}[(20 - 25) + (10 - 10) + (40 - 25) + (30 - 40)] = \frac{1}{2}[5 + 0 + 15 + 10] = 15$

For Region B,

$G = \frac{1}{2}[(10 - 25) + (20 - 10) + (10 - 25) + (60 - 40)] = \frac{1}{2}[15 + 10 + 15 + 20] = 30$

In calculating the coefficient it is not necessary to take account of the + or − answer to each $X - Y$ value; the formula allows for this by halving the final sum of $X - Y$.

Region A has a pattern of manufacturing employment that is more similar to the national pattern than Region B.

GIS: see *Geographic Information System.*

glacial budget refers to the net balance between *accumulation* and *ablation* within a glacier's system. With reference to a temperate glacier:

- if accumulation is greater than ablation, then the snout of the glacier will advance
- if ablation is greater than accumulation, then the snout of the glacier will retreat
- when accumulation and ablation are equal, the snout's position is stationary.

(Note that ice continues to move downhill within the glacier in both of the latter two circumstances.)

glacial control theory: this theory was put forward by Reginald Aldworth Daly to explain the growth of *coral reefs* in relation to fluctuating sea levels. During the glacial periods of the Pleistocene, sea level was lower as a result of *eustatic* changes and sea temperatures would have fallen. As a result any corals would have been killed and erosion would have leveled off existing reefs to produce platforms of dead coral. At the end of the ice age, as temperatures rose, sea level would rise eustatically and the sea would have been warm enough for coral growth. New reefs would be built up on the eroded platforms and would continue to grow upward as sea level continued to rise. Although this theory explains a number of observed facts, it does not fit all situations. Borings through coral reveal thicknesses up to 340 m and seismic evidence suggests that some corals are over 760 m thick. If only glacial control had been involved coral thickness would not exceed 90 m, i.e. the base of the coral would be related to the low glacial sea level. This evidence tends to support the subsidence theory and act against Daly's theory. However, the subsidence theory cannot explain all features and it may be the case that both approaches are applicable to particular coral sites.

glacial diversion: see *diffluence.*

glacial lake: any lake created largely through the actions of a glacier or *ice sheet.* There are a wide variety of such lakes:

- tarn (cirque lake) – a small circular lake located in the base of a *cirque.* It occupies the natural bowl depression of the cirque, although some are dammed behind a ridge of moraine
- proglacial lake – created adjacent to a glacier or ice sheet by *meltwater* issuing from it
- *ribbon lake.*

See also *kettlehole.*

glacial landform: a feature created largely through the actions of a glacier or *ice sheet* (*glacial processes*).

(See also: *arête, cirque, drumlin, erratic, fjord, glacial lake, glacial valley, hanging valley, lodgment, moraine, nunatak, ribbon lake, roche moutonnée, terminal moraine, till, truncated spur.*)

glacial movement refers to the manner in which a glacier or *ice sheet* flows from one area to another. Ice is a relatively rigid substance that behaves more like a plastic as more stress is applied to it. The greater the pressure applied to it, the more it will deform in the direction of that pressure. The movement of ice has two main components:

- basal slippage – the sliding effect of a glacier over the bedrock surface. There is frequently an increase in pressure and friction between the bedrock and the ice, causing a slight rise in temperatures that results in the melting of some ice. The resultant meltwater acts as a lubricant enabling the glacier to move more easily. (See *pressure melting point.*)

- internal flow – movements within a body of ice resulting from the stresses applied by the force of gravity. Such movements often result in tears or crevasses within and at the surface of the ice. These may extend several meters down into an ice mass.

Glacial movement is greatest in temperate areas where more meltwater is available, and in areas with steep gradients. For temperate glaciers, the rate of movement is largely determined by the gradient of the valley in which it is located. Where a glacier moves into a section of the valley that has a reduction in its gradient (i.e. it is flatter), the ice will decelerate and become thicker. This is called compressing flow. Conversely, if the gradient becomes steeper, the glacier will accelerate and become thinner. This is called extending flow.

glacial processes are the actions of a glacier or *ice sheet* either within itself or on the area immediately around it. Such processes include erosion, *glacial movement*, and deposition. The outcome of many processes will be to create distinctive *glacial landforms*.

glacial valley (or glacial trough): a steep-sided, wide and flat-bottomed valley, previously occupied by a *glacier*. The glacier filled the whole valley floor and performed erosion over its entire width. Subsequent deposition of sediment may have flattened the floor even more. The lower parts of the valley sides have often been smoothed by the passage of the glacier, whereas the upper parts are more jagged. The long profiles may be stepped, with rock basins occupied by *ribbon lakes* alternating with slightly raised sections. They are usually straight in their plan, with preexisting interlocking spurs removed by the moving ice to form *truncated spurs*. Other typical glacial landforms such as *roches moutonnées* and *hanging valleys* may be found.

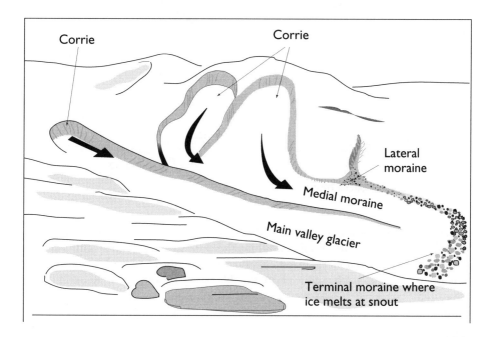

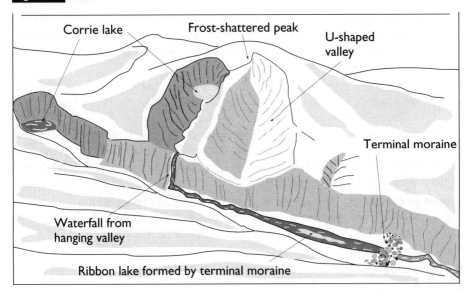

Glacial valley

glacier: a tongue-shaped mass of ice moving slowly down a valley. Processes and landforms associated with such a feature are said to be glacial.

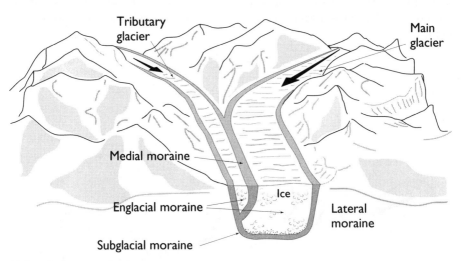

Valley glacier

gleying: a process of soil formation that takes place when the conditions are water-logged or *anaerobic*. Under such conditions the pore spaces are filled with stagnant water that becomes deoxygenized, causing the reddish colored oxidized iron (ferric) to be chemically reduced to the blue-gray ferrous iron. Reoxygenizing causes red-orange patches to appear within the blue-gray soil, this being known as mottling.

global brands are branded products, such as Coca-Cola™, McDonald's™ and Bacardi™, that have been marketed worldwide, the key to which is standardization. In other words, a Big Mac™ should taste the same in Miami, Milan and Melbourne.

globalization refers to the analysis of geographical processes and effects at the global scale. This has become increasingly important because of the large scale of operation of physical, human and economic systems. *Global warming*, the balance between population and resources, and global corporations supplying the global market, all require larger scale analysis because of the interrelationships at this scale.

global shift refers to the locational movement of manufacturing production on the global scale. In the late 19th century and for much of the 20th century manufacturing activity was concentrated in North America and Western Europe, but in the 1970s and 1980s, *transnationals (TNCs)* began to establish plants in the *newly industrialized countries*. The traditional areas became more oriented toward tertiary activity and research and development. The focal point of manufacturing activity has to some extent shifted to the countries around the Pacific Ocean – Japan, Korea, Taiwan, Australia and the west coast of the USA and Canada – the area referred to as the Pacific Rim.

global village: a term used to convey the idea that the world, in terms of transportation cost and time, is "shrinking." Transportation of raw materials and products, transfer of capital, and communication of information and ideas have all become much faster as a result of technological developments. Events taking place in almost any part of the world can be transmitted by satellite and shown on television as they happen, and the emergence of the Internet and the information superhighway further emphasize the trend. This space-time convergence in terms of a global village is to some extent exaggerated: many of the new forms of communication simply follow established networks and therefore access to new technology and the instant world are not universally available.

global warming is the gradual warming of the Earth's atmosphere due largely to man's activities. Evidence for this warming comes from *climatic data* and the *recession of glaciers* and ice margins. The atmospheric concentration of carbon dioxide has increased by about 15% in the last 100 years and the current increase in CO_2 is considered to be approximately 0.3% per year. Carbon dioxide allows incoming shortwave radiation to pass through but it absorbs some of the longwave radiation emitted from the Earth to space. This produces a warming of the atmosphere, the *greenhouse effect*. It is generally agreed that this change will cause a rise in surface temperatures, but it is not yet possible to predict the extent or speed of the increase. A further doubling of CO_2 levels could perhaps raise average surface temperatures by 2–3°C, with greater warming in higher latitudes, perhaps 7–8°C. At current rates of increase this would be reached toward the middle of the 21st century. One of the major reasons for the increase in CO_2 has been the burning of fuels that contain hydrocarbons – coal, natural gas and oil – but large-scale deforestation of areas such as the Amazon Basin have also contributed as trees are a major store of nonatmospheric carbon dioxide. The changes to the Earth's climatic regions and the melting of ice cap areas have serious implications for major areas of food production and low-lying coastal zones, which would be inundated by a rise in sea level.

GNP: see *Gross National Product.*

Gondwanaland is the name given to the former supercontinent comprising the present-day continental areas of South America, Africa, South India, Australasia and Antarctica. Alfred *Wegener* suggested that about 200 million years ago all the continental areas were joined as one land mass that he called *Pangaea*, but in late Mesozoic times the land mass broke up and *continental drift* produced two super-continents, the northern *Laurasia* being separated from Gondwanaland by the Sea of Tethys.

gorge: a deep, steep-sided and narrow valley usually occupied by a river. Gorges are produced when vertical erosion is much faster than lateral erosion. This may result from excessively high river flow as when glacial meltwater discharges through a former valley and undertakes rapid downcutting. It may also occur when an existing river course is interrupted by uplift of land, as in *antecedent drainage* when the river cuts vertically in order to maintain its course. If the downcutting is fast enough the river will continue on its original course through a newly cut gorge.

graben: the downfaulted section of a rift valley system. When extending forces operate in an area of crustal rocks, faults develop and the mass between two parallel faults will move downward. The blocks that are left upstanding on either side are called horsts.

graded profile: the long section of a river that some researchers have suggested displays an even and progressive decrease in gradient down valley. This idea was originally put forward by William M. Davis, who argued that irregularities in the profile, which could reflect changes in underlying geology, are worn away by river erosion to give a smooth graded profile. This is also referred to as the profile of equilibrium and is linked to the concept of a graded river, which has been defined by some geographers as being attained when the river uses up all of its energy in the movement of water and sediment so that no free energy is left to undertake further erosion. Such a balance between energy and work cannot occur at a particular point in time, but is suggested as an average position over a long time. Recent approaches consider rivers as open systems with inputs and outputs of energy, water and sediment. These systems can achieve equilibrium when inputs and outputs are balanced, but changes in the system bring adjustments in the profile as the river attempts to counter the change. In this way it regulates the system. Modern studies

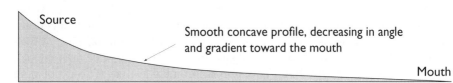

Source

Smooth concave profile, decreasing in angle and gradient toward the mouth

Mouth

Graded profile

have shown that even irregular profiles may be "graded" in that they are in equilibrium but the concave form of long section is very common and can be explained in relation to the controlling factors. The increased efficiency of the channel, with a larger *hydraulic radius* and increased *discharge* downstream, enables the river to transport sediment over a progressively shallower gradient.

gravity model: this can be used to predict the amount of interaction between two places in relation to their distance apart. The predicted interaction is calculated using:

$$\text{FORMULA} \qquad I_{ij} = \frac{P_i \times P_j}{(d_{ij})^k}$$

where P_i and P_j are the populations of the two settlements, d_{ij} is the distance between the two places and k represents a mathematical power, i.e. 1, 1.5, 2, 2.5, 3 etc.

If $k = 1$, the bottom line of the equation would be (d); if $k = 2$ the bottom line would be $(d)^2$ and so on.

A low value of k represents a landscape that has good transportation networks or is easy to cross because it is a flat area.

A high value of k represents an area over which interaction would be more difficult; a high value of k indicates greater resistance to movement.

By predicting interaction between places and comparing this with actual or observed levels of movement it is possible to calculate the value of k. If I, P_i, P_j, and d are known values, then by substitution and rearrangement of the formula it is possible to calculate k. Studies suggest that k normally equals 2 and the formula is usually written in the form:

$$I_{ij} = \frac{P_i \times P_j}{(d_{ij})^2}$$

Interaction should refer to the movement of people, or goods, or the volume of transportation services between places. It could also be applied to the interaction of people in terms of telephone calls between points in the network. Because distance is not always a good indicator of the separation between places, interaction could be predicted using time or cost of movement in the equation.

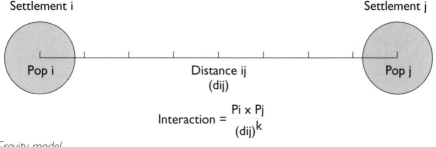

Gravity model

greenfield site: a site for a new factory that has no history of the manufacture of the product in question.

Pros:
- site is chosen on modern rather than historic criteria
- traditional restrictive labor practices will not be present to hinder productivity

Cons:
- local labor may not have right skills or temperament
- local *infrastructure* may need development.

greenhouse effect: see *global warming*.

green movement: the environmental, or green, movement has been well established in the USA for over one hundred years. In recent times it has emerged as a strong political force, with parties such as the Die Grünen (Green) party of Germany taking 8.3% of the vote in the 1987 elections. In the 1989 European elections, green parties received high levels of support, nearly 15% for example in the United Kingdom.

Greenpeace: a green organization established in 1971 by activists protesting about American nuclear tests in the Aleutian Islands. The organization has been particularly involved in the following campaigns:

- controlling the whaling industry
- protecting seals, dolphins and porpoises
- reducing trade in endangered species
- closing nuclear power stations, nuclear reprocessing plants, and stopping the disposal of radioactive waste (and other chemicals) at sea
- stopping tests of nuclear weapons
- stopping acid rain and other atmospheric pollution
- keeping Antarctica free from exploitation.

green revolution: in its narrowest sense this refers to the breakthrough in plant breeding that produced *high-yielding varieties (HYVs)* of grain, but it has come to mean the application of modern farming techniques to *Third World* countries. This package consists of technology (including *fertilizers* and *pesticides*), water control and *mechanization*. The impact of the green revolution has been seen in India when the government in the early 1960s, faced with food shortages, encouraged the introduction of new methods. India has therefore seen the ways in which the green revolution can benefit farmers, but it has also seen some of the drawbacks:

Pros:
- wheat and rice yields have more than doubled (in some areas up to a fourfold increase)
- an extra crop can sometimes be fitted into the year
- farmers' living standards have risen
- farmers can now afford tractors, better seeds, fertilizers and pesticides
- there is now a much more varied diet among rural dwellers
- new jobs have been created in such areas as fertilizer manufacture
- local infrastructures have improved along with markets
- areas under *irrigation* have increased

Cons:
- HYVs require heavy fertilizer application

- HYVs need more weed control, are more susceptible to pests and diseases and are not as well adapted to drought
- some environmental damage is caused through increased use of fertilizers and pesticides
- some farmers have certainly become richer, but many have grown poorer, increasing the differentiation between farmers and therefore disrupting the rural fabric
- this, in turn, has led to more rural-urban migration
- rural indebtedness has increased as farmers borrow money to pay for these inputs
- mechanization has increased rural unemployment
- some irrigation schemes have resulted in *salinization*
- some HYVs have an inferior taste.

green tourism: see *ecotourism*.

groin: a wooden construction built across a beach designed to trap and then stabilize sand and shingle. Sediment brought by *longshore drift* piles up on the updrift side until it is high enough to overtop it, or can drift around the end. They are usually placed at 5° or 10° to the perpendicular in order to prevent scouring on the downdrift side, although the exact angle will depend on the direction of the prevailing waves. Groins can have disastrous effects for areas on their downdrift side. Deprived of their sediment, beaches are actively eroded, allowing waves to attack the cliffs behind. Tourism may also be affected by the loss of beach material.

gross domestic product (GDP) is the sum total of the value of a country's output over the course of a year. It differs from *gross national product* because it does not include net income from abroad.

gross national product (GNP) is calculated by adding the value of all the production of a country plus the net income from abroad. Net income from abroad is the income earned on overseas investments less the income earned by foreigners investing in the domestic economy. In most countries, the growth in real GNP per head of population is the main measure of *economic growth*.

groundwater: water that collects underground in the pore spaces in soil and rock. When it fills all the pore spaces available, the rock or soil is said to be saturated. This water can be transferred slowly through rock as groundwater flow or *baseflow*.

Group of Seven (G7) is the collective name given to the seven richest nations in the world: the USA, Japan, Germany, France, Canada, Italy and the United Kingdom. Because of their economic power they carry considerable political muscle, and consequently their meetings are reported widely. At the Birmingham (England) summit in 1998 Russia fully participated for the first time, giving birth to the G8.

growing season: that part of a year when the average temperatures are above 6°C. It may also refer to the length of time between the last severe frost of spring and the first of the autumn. This is therefore linked to the number of frost-free days that are required for the growth of a particular crop.

growth pole: a relatively small area within a country in which new economic development is targeted. The aim is to increase employment in such an area, and so increase the spending power of the local population. This then creates additional jobs in the service and construction industries as well as attracting more firms linked to the original industry. Growth poles therefore attract migrants, entrepreneurs and capital. Their creation is an integral part of Gunnar *Myrdal's* model of economic development.

Growth poles have been adopted by many governments as part of an effort to reduce regional imbalance. Investment is concentrated in selected towns in the *periphery*, which will then, it is hoped, set off the *cumulative causation* processes. Incentives are often offered to attract industries to them, and away from the *core* region. An example of such a policy was the French government's creation of métropoles d'équilibres to act as counterattractions to Paris. Investment was directed toward cities such as Lyon, Rennes and Grenoble.

One of the main problems with a growth pole policy has been that few industries have relocated themselves entirely to such areas. More often branch plants have been established, which are often the first to close during times of economic recession. Also, it is difficult to determine to what extent growth poles stimulate the local economy. Firms often retain their links with their former areas rather than create new ones.

guest worker (or Gastarbeiter) is a euphemism for an economic migrant from a poorer part of the world who has been attracted to provide cheap labor in an economically more developed country, such as Germany. Such migrants usually take up jobs that are not taken by locals because the jobs are dirty, unskilled and demand long and unsociable hours. Host countries hope that guest workers will only stay for a short period of time, say, two to three years, but in many cases they have not only remained longer, they have brought their families to live with them.

gullying: an extreme form of *soil erosion* producing steep-sided water courses that are subject to *ephemeral flash floods* during rainstorms. The United Nations Food and Agriculture Organization defines a gully as a stream channel more than 0.5 m deep that prevents the normal use of farm machinery. Gullying commonly occurs on steep slopes that have little or no vegetation. Heavy rains are not intercepted and are unable to infiltrate into the ground. Water flows over the surface of the ground as sheet-flow, but small channels or rills are then formed on the surface, which in time develop into gullies.

Gutenberg channel: a layer in the upper part of the Earth's *mantle* in which the speed of earthquake waves is reduced. It is a low velocity channel at varying depths between 100 and 200 km that is made up of less rigid material. The waves from earthquakes originating in this zone take longer to reach a given seismic recording center. The existence of this zone was suspected by the American seismologist Beno Gutenberg as early as 1926, but it took a further 30 years before worldwide recordings of waves from underground nuclear blasts provided confirmation.

guyot: a type of seamount or flat-topped hill rising above the ocean floor. The top of the feature is below sea level and has been planed off by erosion. Many guyots appear to be the remnants of volcanic peaks. They are often referred to as tablemounts.

habitat: the specific environment in which plants and animals live. It may be regarded as the "home" of an *organism* or group of organisms.

Hadley cell: the circulation of air taking place on either side of the thermal equator resulting from convection and subsequent subsidence. Solar heating close to the equator causes masses of warm air to rise creating surface low pressure. (See *inter-tropical convergence zone.*) This air rises to the *tropopause* where it spreads out polewards at high altitudes. Further cooling of this air eventually causes it to subside back to the surface at approximately latitudes 30°N and 30°S to create surface areas of high pressure. Some of this air then returns across the surface to the equator as the *trade winds.* These are deflected by the effect of the *Coriolis Force* to create the northeast and southeast trades in the northern and southern hemispheres respectively.

Hadley cells are to be found on either side of the thermal equator, although their position is not always constant due to movement of the overhead sun. They are named after George Hadley, who in 1735 formulated a model to explain why the so-called trade winds were so reliable for mariners in the tropics.

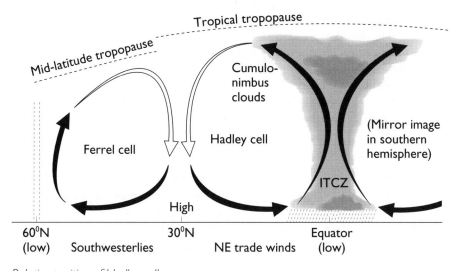

Relative position of Hadley cells

halophyte: a plant that can tolerate saline conditions. An example of a halophyte is sea couch grass, which is a common feature of coastal sand dunes.

halosere: a plant *succession* that takes place in areas where conditions are naturally saline. For example, in a British river estuary bare mud is initially colonized by green algae that can tolerate submergence for most of the 12-hour cycle. This traps further accumulations of mud that are then colonized by *halophytes* such as salicornia and spartina. The latter's long roots allow it to trap even more mud, which forms the habitat for plants less tolerant of inundation and salinity such as sea aster, sea lavender and some grasses. Trapping of sediment is now greatly accelerated, the level of the land is raised and the vegetation cover becomes more dense. A *saltmarsh* is created where the frequency of tidal inundation is reduced. Further accretion tends to be uneven, and so hollows and creeks are produced where the salinity may be too high, or the inundation too frequent for plants to exist. Further inland, juncus and other rushes and reeds grow, eventually being replaced by nonhalophytic shrubs and small trees of rowan and alder.

hanging valley: a tributary glaciated valley perched above the level of the main glaciated valley floor. Streams flowing down these tributary valleys often descend to the main valley floor by waterfalls. They were formed by smaller glaciers joining a main glacier, but the smaller glacier had less erosive power and so could not erode as deeply.

Harris-Ullman model: states that the land use pattern in cities is not built around a single center but around a number of discrete nuclei. It is alternatively known as the *multiple nuclei model.* Some of these nuclei have existed for years, based on villages that have been incorporated into the city's growth. Others have formed more recently, such as new industrial parks acting as foci for suburban residential development. In time there will be an outward growth from each nucleus until they merge as one large urban center. Other main features of the model are:

- similar activities group together because they benefit from agglomeration, for example, office districts
- certain activities are detrimental to others, and so they repel each other, for example, heavy industry and residential areas
- those activities that can afford to pay the highest rates obtain the most desirable sites
- if a city becomes too large, some activities may be dispersed to new nuclei, for example, a new out-of-town mall.

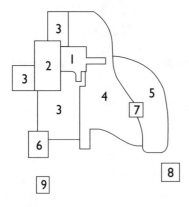

Central business district	1
Wholesale light manufacturing	2
Low-class residential	3
Medium-class residential	4
High-class residential	5
Heavy manufacturing	6
Outlying business district	7
Residential suburb	8
Industrial suburb	9

Harris-Ullman model (multiple nuclei theory)

hazard: an event perceived to be a threat to people, property and nature. It may take place in a variety of contexts:

- in the physical environment, for example, earthquakes and volcanoes
- in the built environment, for example, the disposal of chemical waste
- in the human environment, for example, the spread of a transmittable disease.

The occurrence of a hazard may be the result of changes to environmental conditions that are sudden and difficult to predict in terms of time, duration, location, spatial extent and human impact.

HDI: see *Human Development Index.*

headland: a coastal promontory resulting from the existence of a more resistant rock. A headland usually exists where there is a concentration of erosion by the sea, and erosional landforms such as caves, arches and *stacks* are common features.

heat budget: see *energy budget (the Earth).*

heathlands are areas of rough pasture consisting of coarse grasses and shrubs such as heather. They can be found in a variety of locations:

- wetter upland areas where dwarf shrubs of heather, bilberry and crowberry exist as well as gorse, rushes and sedges. Poorly drained sections within these areas are characterized by sphagnum moss and cotton grass
- drier lowland sites based on fluvioglacial sands and gravels or coarse sandstones. Here purple moor grass and bracken exist alongside heathers.

Many heathlands owe their origin to the interference of humans – they are an example of a *plagioclimax*. The clearance of upland forests over the last few centuries has allowed heather to take advantage of the newly created exposed nutrient-deficient sites. Subsequent heavy grazing and burning have prevented the regeneration of forests and have allowed the heathlands to continue to exist.

heavy industry is the large-scale production of goods that are either the basis for other products, or are large in size. Such industries are very reliant upon the availability of coal and other bulky raw materials. Examples of heavy industry include the iron and steel industry, shipbuilding and chemicals.

helical flow is a downstream spiraling of water flow associated with meandering channels. The main current tends to flow in a straight line and therefore strikes the outer bank of the channel where the river bends. This leads to a return flow of water toward the inner bank close to the bed of the channel. The outward surface flow and inward bed flow produce a circular motion when viewed in cross-section. As the main flow of the channel is downstream, this circular pattern is superimposed, producing a spiral movement or helical flow. This water movement may help in the transport of sediment, moving material from the undercut outer bank toward the inner bend, where it is deposited as a *point bar*. This type of flow is also called helicoidal flow.

HEP: see *hydroelectric power.*

herbicides are chemicals applied to crops to control weeds.

heritage industry: activities that seek to develop landscapes (including buildings), either rural or urban, as a resource. These activities may involve the preservation and conservation of the landscape, as well as presenting and exploiting it for its tourist potential. The marketing of the landscape for its tourist potential is an important element of the heritage industry.

hierarchy: the arrangement of features in an order of importance. In a geographical sense, one of the most significant hierarchies is that of settlements. In simple terms, a hierarchy based on size is: hamlet, village, town, city, conurbation. (See also *Central Place Theory* and *functional hierarchy*.)

Highland Park Plant: the huge factory in Detroit in which Henry Ford achieved the breakthrough to moving, conveyor belt assembly of cars in 1913. This heralded the start of the *mass production* era and therefore the end for many craft manufacturing skills. Prior to the switch to mass production it had taken 750 minutes to assemble a car. By early 1914 it took only 93 minutes, a reduction of 88%.

high-order goods and services have a large trade area and require a large population to support them. They tend to be goods that are purchased less frequently such as furniture and electrical equipment, or services such as accountants or barristers. Customers are willing to travel greater distances to obtain these more specialized high-order functions. These goods have a high *range* and high *threshold*. *Central places*, or settlements, which provide such high-order goods, tend to be further apart in the landscape than service centers, which sell low-order goods.

high yielding varieties (HYV): applies mainly to grain crops where developments in plant breeding produced seeds that could more than double the yields over existing varieties. Initial development took place in Mexico and later at the International Rice Research Institute (IRRI) in the Philippines where the so-called "miracle rice," IR-8, was produced. Recent work, particularly in Malaysia, has seen a whole range of "designer" seeds produced for individual farmers' needs. From work on wheat and rice, research has now gone into other crops such as maize and potatoes. HYVs therefore formed one of the major developments that was part of the "green revolution."

Pros: • can give up to four times the yield of former varieties
      • some are more responsive to *fertilizers*
      • shorter growing period allows double or triple cropping to occur
      • many varieties can be grown in a range of environments

Cons: • require heavy doses of fertilizer
      • less adaptable to drought
      • many varieties are more susceptible to pests and diseases, therefore requiring large inputs of *pesticides* and *fungicides*
      • need more careful weed control
      • new seeds have to be bought each year to ensure the purity of the strain
      • some varieties have an inferior taste.

Hirschman, Albert O.: an American economist who established some of the principles of the *core-periphery* model or the theory of *polarized* growth. Hirschman maintained that the movements of people, *entrepreneurs* and *capital* to the core was a process of polarization. Growth is then supposed to trickle down to the periphery in order for it to develop. The demand from the core for food and resources from the periphery, for example, should stimulate economic activity in that area.

histogram: a method of illustrating a *frequency distribution* consisting of a set of vertical bars that rise from a graph's (horizontal) x-axis. The (vertical) y-axis is labeled "frequency." The data is arranged into classes before plotting.

Worked example

Input of fertilizer per hectare on sixty farms

2	5	8	13	17	22
2	6	9	13	17	23
2	6	9	13	18	25
3	6	9	14	18	26
3	6	10	14	18	26
3	7	10	14	18	27
4	7	11	14	19	28
4	7	12	16	20	31
4	8	12	16	20	32
5	8	12	17	21	36

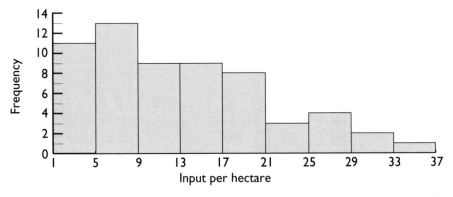

Histogram

hi-tech industry: the term high-technology industry refers to those that have developed within the last 20–30 years and whose processes often involve micro-electronics. Also included are industries such as *biotechnology*, pharmaceuticals and the manufacture of medical equipment. It could also be defined as industry that has high inputs of scientific *research and development* and produces new, innovative and technologically advanced products.

Hjulstrom curve: a graph that shows the relationship between stream velocity and the size of sediment that is picked up, transported and deposited. Velocity increases as discharge rises, enabling the stream to pick up particles from the bed or banks of the channel. Filip Hjulstrom's research showed that lower pick-up or critical erosion velocities were needed to move sand particles than for finer silts and clays or coarser gravels. The small-sized clay particles are difficult to pick up because of their cohesive properties; they do not occur as single particles. Silts and clays also lie on the channel floor; because of their smaller size they offer less resistance to water flow than larger particles and therefore a more powerful flow is needed to lift them. Once picked up, particles can be transported at velocities lower than the critical erosion velocity. As discharge and velocity decrease, sediment is deposited following the sequence shown by the fall line or settling velocity curve. Larger grains are deposited first and there is a progressive reduction in grain size as settling velocity decreases. Clays and smaller silt particles, once picked up, are only deposited at very low velocities; this explains the transportation of these fine materials into the lower reaches and estuaries of rivers where they may form mud flats.

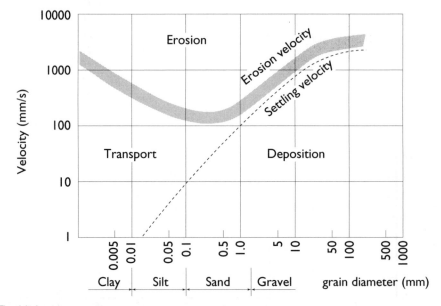

The Hjulstrom curve

Holocene: the geological period covering the last 10,000 years since the end of the *Pleistocene* glaciation. Within this period there have been a number of major fluctuations in climate in North America and Europe that have resulted in significant vegetation changes.

homeworking: earning income from work undertaken at home. Traditionally homeworkers have completed labor-intensive, very low-paid jobs such as hand sewing and packing. Modern technology offers the possibility that more professional employees could work from home, armed with communication links such as electronic mail, fax and telephone.

honeypot: places with special interest or appeal that become very popular with tourists and tend to become overcrowded at peak times. This can give rise to a number of problems associated with overuse, particularly with congestion on access routes. In some popular areas it has become necessary to limit vehicular access at certain times such as in Yellowstone National Park. Where planners wish to concentrate tourists in order to protect other areas, it is not unusual for honeypot sites to be developed, with adequate visitor amenities (toilets, parking lots and picnic facilities).

horizontal integration occurs when a firm takes over or merges with another firm at the same stage of production. The production process starts with raw materials being processed, then moves through manufacture and assembly, followed by sales to wholesalers and then retailers. Horizontal integration can occur at any of these stages. The merger of two breweries would be an example of horizontal integration and would have the following advantages and disadvantages:

Pros:
- increases market power over the next or the previous link in the process
- enables greater *economies of scale* to occur

Cons:
- may restrict customer choice
- could be seen as a *monopoly* and thus be referred to the Federal Trade Commission or the Department of Justice Antitrust Division.

horticulture: strictly speaking, the art of gardening, but in practice it is taken to include the production of fruit, vegetables and "ornamentals" (decorative non-edible plants, e.g. flowers). It can be carried out both in the open and under protected structures such as greenhouses. Horticulture is highly *intensive agriculture*, particularly in terms of labor, usually carried out on small plots with large investments of capital for technology. Shortages of labor have led to some farms developing as Pick Your Own establishments.

hot spots: areas where heat under the Earth's crust is localized. At such points, rising *magma* can produce *volcanoes*. The Hawaiian Islands, whose summits are volcanoes, have been formed in this way.

Hoyt, Homer put forward a theory on the internal structure of cities that was based on sectors rather than the concentric zones of the *Burgess* model. His original work in 1939 was based on studies in 142 American cities from which he suggested that the key determinant of housing patterns in cities was the choice of residential location made by the wealthy who could afford the highest rents, which were often along major roads. He also observed that high and low rental areas repelled each other.

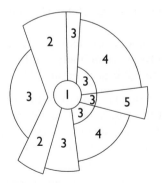

1 CBD
2 Wholesale, light manufacturing
3 Low-class residential
4 Medium-class residential
5 High-class residential

The Hoyt sector model

Huff, Darrell L.: developed a probability model in 1962, which attempted to predict the probability of people purchasing goods in different shopping centers. The model takes into account the relative attraction of different shopping centers in an area and the distance that would have to be traveled in each case. Huff maintains that shoppers consider distance and attractiveness of towns. The attraction of towns can be measured in terms of such variables as population size, number of shops, number of types of shops and total turnover.

FORMULA

$$\text{Probability of visiting town A} = \frac{\dfrac{\text{attraction of town A}}{\text{distance to town A}}}{\dfrac{\text{total attraction of all towns under study}}{\text{total distance to all towns under study}}}$$

human development index (HDI): a modern method of trying to define development by using several indicators. This was first proposed by the United Nations in 1990 and uses three variables based on income, education and *longevity*:

- income is measured by adjusting it to the purchasing power within the country
- education is taken from the adult *literacy* rate and the average number of years spent in school
- longevity is taken as the *life expectancy* at birth.

For each variable, a country scores according to its position on the list on a scale from one to zero. If a country is at the top of a list its score will be 1. If it is at the top of all three lists, its combined score would be 3, which averages out to an index of 1. On this basis, Japan is the most developed country in the world with an HDI of 0.993. The country showing the least development is Sierra Leone with an index of 0.048. The HDI also gives an idea of how a country rates in terms of the best conditions. Honduras, with an index of 0.492, could be said to be about half as developed as Japan.

There are, however, a number of criticisms of the HDI:

- quantity says nothing about quality – what standards have been reached, for example, in literacy
- the index is a measure of relative rather than absolute development through its use of rankings
- it is not that far removed from classifying development according to *GDP/GNP*, as all the variables depend to a large extent upon wealth.
- it contains no measure that refers to human rights or freedom.

humidity is the amount of gaseous water (or water vapor) content of the atmosphere. This content varies a great deal throughout the atmosphere, being virtually zero in polar regions to over 5% in parts of the tropics. The amount of water vapor in the atmosphere is dependent on temperature: the warmer the air, the more water vapor it can hold. Humidity can be measured in three ways:

- absolute humidity – the total amount of water vapor in a given mass of air, usually given as grams per cubic meter
- vapor pressure – as above, but in this case expressed as the amount of pressure exerted by the water vapor content compared with the atmospheric pressure as a whole, e.g. 20 millibars out of a total pressure of 1000 millibars

- relative humidity – the most important of the measures, being defined as the actual vapor content in comparison with the amount that the air could hold at the given temperature and pressure, measured as a percentage.

humus: the organic matter in the soil that is derived from the breakdown of plant and animal matter and secretions of living organisms. To be classified as humus this material has to be completely broken down. Humus gives the soil a dark color and in combination with clay particles (the clay-humus complex) is one of the major contributors to *soil fertility* as it provides it with a high water and nutrient-holding capacity. It also helps bind the soil together and thus combats *soil erosion.* There are two forms of humus, which vary in *acidity*: *mull,* a mild form of humus and *mor,* the acid variety. Some soil scientists recognize an intermediate category known as moder.

hurricane: the name given to a tropical *cyclone* that occurs in the Atlantic and Caribbean areas. Hurricanes are systems of intense low pressure (up to about 600–700 km across), formed over tropical sea areas and moving erratically westward until they reach land, where their energy is rapidly dissipated. At their centers they have an area of subsiding air with calm conditions, clear skies and higher temperatures, known as the eye. A number of factors encourage the development of hurricanes:

- the heat energy and abundant moisture supply of tropical sea areas
- the release of vast amounts of *latent heat* by *condensation* once *convection* has begun
- the deep layer of humid and unstable air to be found on the western sides of ocean basins
- a circulatory motion of the air, counterclockwise in the northern hemisphere (encouraged by the *Coriolis Force*).

The major *hazards* associated with hurricanes are:

- strong wind speeds often over 200 km per hour
- heavy rainfall with associated flooding – e.g. 40 cm in 24 hours
- *storm surges* – hurricane Camille (1969) had a 7 m surge on the Gulf coast
- *landslides* after heavy rain.

The cost of a hurricane, in both human and economic terms, can be enormous. In *First World* countries such as the USA, there may be adequate warning of the event and preparation for it. Consequently, there often tends to be enormous damage but very little loss of life. Communities are able to adjust and recover in a relatively short space of time. In *Third World* countries, the loss of life, the damage and the economic hardship can be great, particularly when the economy is dependent on a narrow range of agricultural activities.

hybrid: a plant resulting from a cross between two genetically different plants. Many economically important cultivated plants have resulted through natural hybridization or hybridization induced through biological research. Examples include bananas, coffee, peanuts, alfalfa and roses. The process is important for agriculture as hybridization can produce varieties with certain qualities, such as disease and drought resistance, quicker ripening and *higher yielding varieties.*

hydration is a process of chemical *weathering* in which certain types of minerals take up water and expand. This creates physical stresses within the rock that can lead to disintegration and assist other forms of weathering. Calcium sulphate can occur in both an unhydrated state as anhydrite and a hydrated state as gypsum. In chemical terms:

$$CaSO_4 \quad + \quad H_2O \quad \rightarrow \quad CaSO_4H_2O$$
$$\text{anhydrite} \quad + \quad \text{water} \quad \rightarrow \quad \text{gypsum}$$

The uptake of water alters the solubility of the mineral as gypsum dissolves more rapidly than anhydrite.

hydraulic action is a process of stream erosion that moves loose, unconsolidated material due to the force of moving water and its frictional drag on sediment lying on the bed of the channel. As velocity increases, turbulent flow lifts a larger number of grains, particularly sand-sized particles, from the floor of the channel. Hydraulic action is particularly effective at removing loose material in the banks; this can lead to undercutting and collapse.

hydraulic radius: a measure of the efficiency of a channel; it is calculated using the formula:

$$\text{Hydraulic radius (R)} = \frac{A}{P}$$

where A is the cross-sectional area of the channel, i.e. width × depth, and P is the wetted perimeter, i.e. the distance along the bed and banks that is in contact with the water.

A stream loses energy as a result of friction between the water and the bed and banks of the channel. Therefore a stream channel is more efficient, i.e. it loses proportionally less energy in overcoming friction compared with the discharge through the cross-section if the ratio between the area and the wetted perimeter is high. As channels become larger they tend to be more efficient; area increases at a faster rate than wetted perimeter.

For example, a channel that is 10 m wide × 2 m deep has a value for R (if at bank full) of $20/14 = 1.43$. A channel that is 20 m wide × 4 m deep has $R = 80/28 = 2.86$. The area has increased fourfold but the wetted perimeter has only doubled.

hydroelectric power is a clean and efficient renewable resource generated by using the power of running water to turn turbines. This can occur at natural waterfalls, but more usually involves the construction of a dam to impound a lake. This creates a head of water, determined by the height difference between the water intake pipe and the turbine room, which produces the pressure required. Conditions favoring HEP schemes are stable underlying geology, impermeable drainage basin, high and reliable rainfall, and mountainous areas with narrow gorge-like valleys. Access to markets is also important although some activities such as aluminum smelting, electrochemical and metallurgical industries, which have very high electricity consumption, may be attracted to the HEP site even if it is remote. Although HEP is 80–90% efficient (based on the proportion of input that can be used as energy), compared with oil and coal at 30%, capital costs are high and schemes are expensive to install. Many of the most suitable sites are often remote from markets, although recent improvements in technology allow transmission over distances in excess of 800 km. Despite its great potential it only contributes a small proportion of total energy, but it is an important source in parts of the world with few oil, gas and coal reserves, such as South America, Africa and parts of Southeast Asia. It is a major power resource for some developed countries such as Norway, Sweden and Switzerland.

hydrograph: a graph that shows variations in river *discharge*, in cubic meters per second, over a period of time. They can be plotted for a period ranging from a few hours to several months or even for a year. A storm hydrograph is plotted after a storm event to record the effect of the input of rainfall on the discharge in the river system. The starting and finishing level of discharge is called the base flow, representing the river flow produced by groundwater seeping into the channel bed. As stormwater enters the system, the discharge rises, shown by the rising limb, to reach peak discharge, which indicates the highest flow in the channel. The recession limb shows the fall in discharge to a final base level. The time delay between maximum rainfall intensity and peak discharge is called the lag time. The rising limb is produced mainly by overland flow, which reaches the channel quickly; the steepness of the limb depends on the intensity and duration of the storm and the amount of antecedent soil moisture reflecting recent weather conditions. Heavy rain falling on a soil that is saturated as a result of a period of wet weather will produce a steep

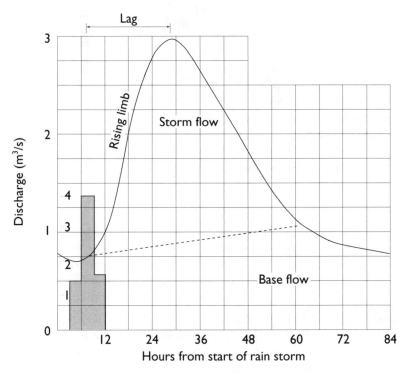

Hydrograph

rising limb. The recession limb is produced mainly by water flowing through the soil into the channel and from overland flow from points at a greater distance from the channel. A small drainage basin tends to respond more rapidly to a storm than a larger basin and a short lag time and steep rising limb tend to follow after a period of very wet weather that has saturated the soil. These quick response hydrographs are described as flashy. The shape of the graph reflects the amount, intensity and duration of the rainfall, the seasonal weather conditions and the level of residual moisture in the soil layer, vegetation cover, the size of the drainage basin and the slope angles within the basin.

hydrological cycle: the circulatory system by which water is transferred between the oceans and the land masses.

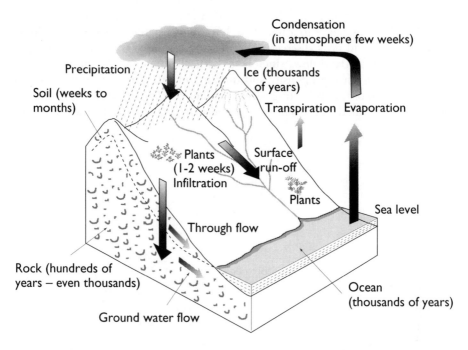

Hydrological cycle

Evaporation from the sea, lakes and the land surface and *transpiration* from vegetation produces water vapor in the atmosphere. As this moist air rises or moves laterally to cross land areas, it is cooled and the resulting condensation produces precipitation, which falls on the land and sea. Surface run-off in streams or rivers and percolation that supplies *groundwater* return water to the sea.

hydrolosis is a type of chemical *weathering* in which there is a chemical reaction between a mineral and water. There is a reaction between the H^+ (hydrogen) ions or OH^- (hydroxyl) ions in the water and the ions of the mineral. It is particularly effective in the breakdown of feldspar, which is commonly found in igneous rocks such as granite. The potassium feldspar (orthoclase) is broken down into a clay

mineral, kaolinite, together with soluble potassium oxide and a soluble silica, or sili-
cic acid, which are both removed in solution. This leaves a residual clay deposit that
is considerably less resistant than the original granite. In chemical terms:

FORMULA $2KAlSi_3O_8 + 2H_2O \rightarrow Al_2Si_2O_5(OH)_4 + K_2O + 4SiO_2$

orthoclase   water   kaolinite   potassium soluble silica
oxide

hydrosere a type of *primary succession* that begins on a waterlogged or submerged
site. These wet environment *vegetation successions* may be divided into hydroseres,
which develop in freshwater sites such as lakes or ponds, and *haloseres*, which origi-
nate in saltwater sites such as tidal mud flats. In a pond the first plants to develop
are submerged aquatics. These help to trap sediment that enables other species,
such as aquatics with floating leaves, to move into the area. The next seral stage sees
the growth of reed beds and swamp conditions and, as debris accumulates with
increased thickness of silt and sediment, the surface rises above the water level to
produce a carr or fen. This leads to colonization by alders and ferns that then
further modify the environmental conditions to allow the entry of willows and
ash. Eventually the climatic *climax vegetation* of deciduous oak or beech woodland
is reached. Throughout the succession there are progressive changes to the soil
conditions, the ground level microclimate and the faunal activity as the *ecosystem* is
modified through time.

hygroscopic is a term used to describe a substance that will attract water.
Impurities in the *atmosphere* act as hygroscopic nuclei, allowing *condensation* to take
place. Condensation can occur on salt nuclei at *relative humidities* as low as 78%.
Large amounts of such condensation nuclei are present in the atmosphere and consist mainly of solid particles of varying sizes. The smallest particles are largely produced by combustion, natural and man-made, and these occur in high
concentrations near industrial areas and in polluted urban air. Larger nuclei consist
of ammonium sulfate and sulfuric acid, which are both hygroscopic. These again
tend to be the result of industrial processes or power production and are more
prevalent in urban industrial areas. The largest nuclei are sea salt particles; these
contribute only a small percentage of the nuclei involved in cloud formation.

hypabyssal: a name given to a group of igneous intrusive rocks that cooled and
crystallized at intermediate depths in minor intrusions such as *sills* and *dykes*. These
rocks have grain sizes lying between the coarse size of granite that cooled slowly at
depth and the fine-grained basalt that cooled quickly as a result of lava extrusion
onto the surface. Individual features often display a variety of grain sizes within
one rock type and this group of rocks overlaps with those that are coarse or fine-
grained. In addition to medium-grained rocks such as dolerite, common in dykes
and sills, this group includes volcanic pitchstones and the plutonic granite
pegmatites.

hypothesis: a stated proposition that suggests a relationship between variables.
In statistical terms it is something that needs to be tested. Hypothesis testing
involves expressing an idea about that relationship in order to provide the basis for
the collection of data that can be analyzed statistically to prove or disprove the
relationship. A hypothesis relating to urban development could be that "building
height decreases with increasing distance from the city center" or in a drainage

basin "discharge increases with distance from the source." It is more usual to formulate a *null hypothesis*: "there is no variation in building height as distance from the city center increases." If the result rejects the null hypothesis then there is some significant relationship between the two variables.

HYV: see *high yielding variety.*

I

Ice Ages: the common term for the periods when there were major cold phases, known as glacials, and ice sheets covered large areas of the world. Prior to 1950 it was believed that there were four major glacials during the last two million years, but recent studies of sediments on the ocean floors now suggest that there have been multiple glaciations, each interspersed by warmer phases or *interglacials.*

During the glacials the ice caps of the world expanded greatly. The reason for the onset of the ice ages is still the subject of debate. One theory is that there are celestial cycles when changes to the Earth's position in space occur, to both its tilt and its orbit around the sun. These changes combine to create a reduction in insolation. Another theory is that they are related to variations in sunspot activity.

iceberg: a floating mass of fresh water found in the world's cold seas in or near to polar latitudes. They are formed by:

- blocks of ice breaking off from floating polar ice sheets during the warmer spring weather
- blocks of ice calving away from glaciers as they flow into the sea.

Nine-tenths of any iceberg can be found below the level of the sea, and it is this that makes them such a danger to shipping.

ice cap: a large area of ice burying the landscape of up to 50,000 km² of land. Like the larger *ice sheet,* they tend to have a flattened dome-like cross-section.

ice core: obtained by drilling down through an *ice sheet* in order to examine the rings of annual accumulation of snow. From such rings it is possible to identify:

- the overall age of the ice sheet
- the variations in the degree of volcanic deposition during the timescale of the core
- the varying chemical composition of air bubbles at different stages in the past in order to identify changes in global temperature during the timescale of the core.

Ice core can also be applied to the block of ice found in the center of a *pingo.*

ice lens: an accumulation of ice beneath the surface in *periglacial* and tundra areas that causes the ground to dome slightly upward. Ice lenses are formed in fine-grained soils such as clays and silts. With the onset of winter, the layer of the ground nearest the surface freezes first. The ice in the frozen layer attracts water from below by *capillary action.* A layer of ice begins to grow while freezing continues and unfrozen groundwater is allowed to move upward. The ice layer forms a lens shape parallel to the ground surface, and the thickening of the ice at its center causes the ground to heave upward. Stones resting on the ground surface are rolled sideways by the heaving ground, helping to create landforms such as *stone circles* and other types of *patterned ground.*

ice marginal landform: a feature located and formed at the edge of an ice sheet or glacier. It may include: proglacial lake (see *glacial lakes*), lateral moraine (see *moraine*), *terminal moraine, kame, esker* and *outwash*.

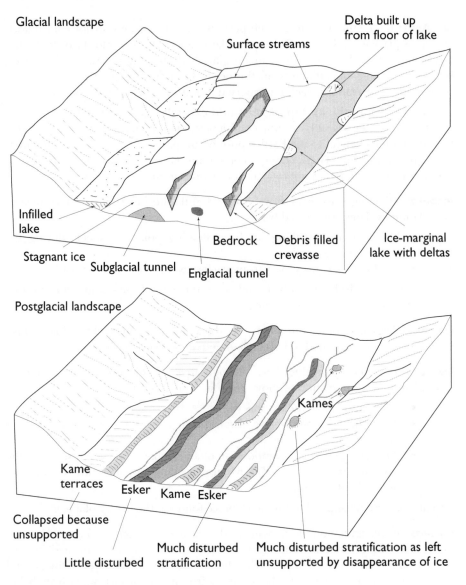

Ice marginal landforms

ice sheet: a body of ice covering the landscape of a very large area of land, over 50,000 km^2. It has a flattened dome-like cross-section, and is hundreds of kilometers in width. The two best known ice sheets are on Antarctica and Greenland.

ice wedge: a deep crack in the ground surface in *periglacial* environments, which is broader at the top and may taper down to depths of 3 m or more. In plan, they frequently form a polygonal pattern across an area. In such areas intense cold temperatures (below $-15°C$) cause a shrinkage in the volume of the ice held within the soil. This leads to the development of cracks in the frozen ground. Early in the following spring, moisture collects in these cracks and freezes in their base since the ground is still frozen at depth (*permafrost*). This prevents the crack from closing again when the ground expands as the temperature rises.

Later more moist times will cause the wedge to be filled with stones and gravels. In this way fossilized ice wedges can often be seen in gravel pits in areas that have previously been periglacial.

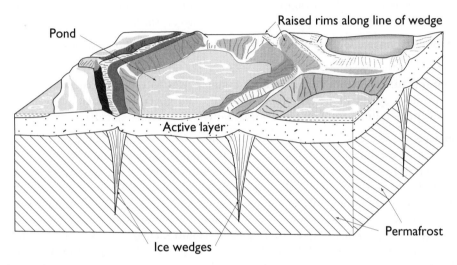

Ice wedges

igneous: a type of rock composed of crystal minerals that forms from the cooling and solidification of molten *magma*. If cooling takes place rapidly, say on the surface, then fine-grained and glassy rocks form, such as basalt. If cooling is slow, usually at depth, then coarse-grained rocks are created such as granite. They are often very hard, rarely show any layering and never contain fossils.

illuviation: the washing in and deposition of material from an upper horizon of a *soil profile* to a lower horizon by downward percolating water. Illuvial horizons tend to be enriched by materials such as clays and organic substances.

IMF: see *International Monetary Fund.*

immature soils mainly reflect the properties and characteristics of the underlying parent material, because soil processes have not operated for sufficient time for the soil to be in equilibrium with its environment, as is the case with a mature soil. An immature soil is still in the process of adjustment to conditions and has poorly developed horizons, a high mineral content and a low percentage of organic material. A newly built-up surface such as a sand dune would develop an aeolian regosol, strongly influenced by the parent material. On steep slopes, where there may be downslope movement of the thinly weathered surface debris, soils such as lithosols develop. Processes are interrupted by the disturbance to the limited soil profile.

immigration is the movement of people into a country, as distinct from *emigration*, which is the movement of people out of a country.

impermeable describes a rock that neither absorbs water nor allows it to pass through. Granite is a good example. Such surfaces produce more *run-off* and therefore a greater number of streams.

import controls are *tariffs* or *quotas* designed to limit the number of overseas goods entering the domestic market. In extreme cases, some goods are prevented from entering at all, which is called an *embargo*. Other forms of import control are collectively called nontariff barriers, which discriminate against imported goods in a more subtle way. One example of such a nontariff barrier would be requiring excessive paperwork for imports, resulting in delays at frontiers.

import penetration: a measurement of the share of the home market taken by importers.

import substitution: this is when a country tries to establish its own industry to produce goods that were once imported. The government then protects the developing industry from foreign competition by imposing *tariff* barriers.

incised streams or rivers have cut deeply into the landscape. The nature of the landforms produced by incision depends to some extent upon the rate of vertical erosion. When incision is slow, and lateral erosion is also taking place, an ingrown meander is produced. The valley becomes asymmetrical with steep cliffs on the outer bend and more gentle slip-off slopes on the inner bends. With rapid incision, where downcutting dominates, the valley is more symmetrical, with steep sides and a *gorge*-like appearance. These are described as entrenched meanders. Where incision is due to a relative fall in sea level, giving *rejuvenation* along the river course, *knickpoints* and *river terraces* will also be formed.

income at the personal level can be seen as wages and salaries. At the level of an economic organization it is the revenue received when goods or services are sold to a customer, which may be an individual or another firm.

independent variable: when investigating the degree of *correlation* between two sets of data, changes in the independent variable are not related to the changes in the other variable (the *dependent variable*). For example, in a study of rainfall total and altitude, altitude is the independent variable: changes in altitude do not depend upon rainfall total. Rainfall is the dependent variable: it is assumed that rainfall will vary with altitude. When plotting a *scatter graph* to investigate the relationship between two variables, the independent variable is plotted on the x (horizontal) axis.

index numbers are convenient ways of showing change in a set of data over time. For example, let us take a car plant that produced 250,000 cars in 1985, 300,000 in 1990 and 350,000 in 1995. If we make 1985 the *base year* and let the value for that year be equal to 100, then the other values will be set relative to this. Therefore, in 1990 the index would be 120 [= $100 \times (300,000 \div 250,000)$] and in 1995 it would be 140. The index tells us quickly that in 1990 output had risen 20% compared with 1985 and in 1995 by 40% compared with 1985.

Index numbers are useful for displaying trends in that easy comparisons can be made between different time periods and between countries with very different absolute values for the variable. However, in using index numbers absolute changes are hidden or ignored; they only display relative or percentage change from the base point.

index of dissimilarity is a statistical measure of the degree of difference between any two sets of paired percentages, for example, comparing the origin of imported goods into the USA at two different time periods. The index is calculated using the formula

$$\Sigma(a - b)$$

where a and b are the pairs of percentages to be compared. So, imports for the first date would be listed as "a," and imports for the later date as "b" for each area of origin. Calculate (a − b) for each pair of percentages, i.e. the percentages for each region of origin and add these together, ignoring the + or − signs.

Providing both sets of percentage data add up to 100, it does not matter whether a is greater than b or b is greater than a, for the same result occurs. The index ranges from 0, perfect similarity (a − b for each set is 0), to 100, where the percentages are as dissimilar as is possible.

The calculation of a single numerical index allows a more objective comparison of data.

index of similarity is a statistical measure of the degree of relative spatial concentration or dispersal of a particular activity or element of the population. It compares two data sets in terms of their spatial distribution. For example, the percentage of an urban area's retired population living in each ward of the city could be compared with the percentage of the total population living in each area. This could indicate the relative concentration of retired people, perhaps highlighting areas of need for particular services. The degree of similarity between the two sets of data can be calculated using the formula

$$Index = \frac{\Sigma d}{100}$$

where d is the difference between the two percentages for each of the city wards. In adding up Σd, use only the + or the − values, as they should be equal. The index can range from 0, where both sets of percentages are identical (each d value = 0, therefore $\Sigma d = 0$) to 1. A low index indicates that the distribution of retired population is similar to the distribution of the whole population, retired people being neither over- nor underrepresented in each ward. A high index indicates that the two data sets are less similar; there are areas that are overrepresented in terms of the percentage of retired population. It is not simply that there are higher percentages of retired population in some wards: they are at a greater concentration in relation to the distribution of the total population, and are overrepresented in some areas.

indigenous means belonging to that area or region, such as resources found within a national territory, plants that are native to a particular environment or ecosystem, or the native population.

industrial inertia occurs when an industry remains concentrated in an area even though the original factors that caused it to be located there no longer apply. Fixed investment in factory buildings and machinery, together with the general advantages of infrastructure in the area (housing, roads, railroads, etc.), have retained the industry. This will continue to be so for as long as the industry remains profitable.

industrial location theory attempts to explain the operation of geographical factors and the decision taken on the location of individual firms and industries. Alfred *Weber*'s theory, published in 1909, emphasizes the role of raw materials, markets and transportation costs in determining the best or optimal location. Because of the assumptions made in the model relating to the decision-making process the optimal point was the *least cost location*. This was achieved when the combined costs of transporting the raw materials to the plant and the product to the market were at their lowest. Weight- or bulk-gaining processes were generally located near the market while bulk-reducing industries were located near, or at, the location of the raw material that lost the greatest proportion of its weight during manufacture. Weber devised the *material index* to indicate the industry's preference for a market or raw material location. Because Weber had viewed the market as a single point in the landscape and emphasized costs as a factor, August *Losch* explored the idea of maximizing profits rather than minimizing costs and focused on revenues from a market area. Revenue would fall with distance from the point of production because of the cost of transporting the goods to the customer. In this way the edges of the market area could be defined where revenue was zero. Both Weber and Losch make underlying assumptions about the behavior of decision-makers (economic man) that limit the application of the theories. The market-oriented approach seems to be more relevant to weight-gaining industries such as brewing, bakeries and assembly activities such as furniture, while least cost approaches are suited to the more traditional heavy industry such as iron and steel. Smith's theory of *spatial margins to profitability* takes into account both revenue and costs, and defines areas in which a firm can make a profit. Individual firms may locate in a suboptimal location where they make a profit, but not the optimum profit. The exact location may reflect other factors: availability of knowledge and ability to use this information, and the aims of the decision-makers. Instead of assuming that all decision-makers are *optimizers*, i.e. they aim to gain maximum return in any situation, it accepts that they may be *satisficers*. They may be satisfied with a less than optimum profit in return for other benefits such as leisure time, or a pleasant environment in which to enjoy family life. This approach was summarized by Allan Pred's work on the behavioral matrix. Because of the element of uncertainty regarding the outcome of such decisions, this is referred to as a probabilistic theory; earlier ideas by Weber and Losch were deterministic, always giving a fixed outcome reflecting the particular factors.

industrial revolution (new) refers to the growth of new industries and businesses that are often based on high technology and that can be developed in a wide variety of locations. There has been a change from the traditional, single item, mass-produced output from a largely unionized workforce (*Fordist* industry) to a *flexible system of manufacturing* with a more varied range of products within a company that can respond quickly to market needs.

industrial types: a classification of the main sectors of industry:

- *primary sector*
- *secondary sector*
- *tertiary sector*
- *quaternary sector.*

infant mortality is the number of deaths of children under the age of one year expressed per thousand live births per year.

infield-outfield: a system of farming in which the small, enclosed fields near the farmhouse (the infield) were heavily manured and cropped continuously while the areas further away (the outfield) were extensively grazed, cultivated at lower intensity with crops that required less attention, or left fallow for long periods. Although at one time this was common in peasant agricultural areas in Europe, it has been largely superseded. Some present-day systems of subsistence cash cropping in West Africa, Bangladesh and Southeast Asia have a similar arrangement around the village, which acts as a source of labor and the market for the crop, with roughly circular patterns of land use decreasing in intensity as distance increases.

infiltration is the passage of water into a soil. Water is drawn into the soil by gravity and capillary attraction. Infiltration takes place at a higher rate at the start of a rainstorm, but as the soil becomes more saturated the infiltration rate falls steadily. Infiltration rates are influenced by:

- how much water is in the soil before a storm, which in turn is a reflection of the rainfall preceding a storm. A very wet soil has a lower infiltration rate than a slightly damp soil. However, a very dry, bare soil tends to have a low infiltration rate since small particles block the main water passageways, and water often rests on the surface
- the texture of the soil – sandy soils have a greater ability to allow water to pass through them, whereas clay soils have much lower infiltration capacities
- vegetation – plants promote a thicker soil and freer draining texture. They also have roots that enable water to pass into and through a soil more easily.

informal sector refers to the form of employment commonly found in the cities of economically less developed countries where people have had to find work for themselves, often illegally, on the streets and in small workshops. It is not part of organized industry, but does employ many thousands of people in unskilled and semiskilled jobs. Some characteristics of the informal sector are:

- self-employed, small family enterprises based on little capital investment
- hours are irregular, and wages are uncertain as bartering is the norm
- many jobs involve selling goods or providing services on the streets, or at small roadside stalls
- raw materials are cheap, and often recycled waste, and so the quality of the goods is usually of a low standard
- a significant proportion of the workforce are children attempting to supplement the family income.

information technology is the means by which information can be stored, processed and communicated almost instantaneously to any part of the world. Advances in computers, telecommunications and microelectronics have allowed this phenomenon to occur, and the pace at which they have taken place has been remarkable. The knowledge of and access to "information" is now of crucial importance in a whole range of activities, and has had a number of geographical effects:

- *homeworking* is becoming more widespread. People can earn an income from work undertaken at home, and can communicate with employers via electronic systems. This has caused some decentralization of employment
- in-home shopping is already established in the USA, and may soon spread to Western European countries. Telephone banking and other financial services are also available
- video-conferencing has reduced the need for people to travel within and between countries. It also reduces the expense of maintaining downtown office locations if there is less need for business people to meet face to face.

The adoption of the new developments in information technology has largely taken place in preexisting core regions, and also in transnational corporations, thus reinforcing their economic superiority. Consequently, although technology offers the chance to decentralize and disperse economic activity, it seems to be concentrating it even more.

infrared is that part of the sun's radiation (*insolation*) that has a wavelength greater than the visible red rays. The wavelengths are those greater than 0.75 micron (0.75×10^{-6} m). Infrared rays penetrate hazy skies and darkness. Hence landscapes obscured by cloud can be photographed using infrared-sensitive photographic plates.

infrastructure is the name given to the road, rail and air links, sewage and telephone systems and other basic utilities that provide a network benefiting business and the community. One of the main advantages that industrialized countries have over less developed ones is the existence of an efficient infrastructure. Building up such a system is very expensive, requiring a great deal of *capital* to be set aside, capital that poorer countries find difficult to afford.

inheritance laws have an effect on farming in some societies if, on the death of the farmer, the land has to be divided equally among the sons. This often leads to the breakdown of the farm into numerous scattered small fields, a process known as *fragmentation*. The tradition of equal division was known as *gavelkind*, as distinct from the situation in which the eldest inherits (*primogeniture*).

inner city: in most cases the inner city refers to those areas of older residential and industrial development lying between the *Central Business District* and the suburbs of major cities and *conurbations*. In the *Burgess* land use model it corresponds to the zone in transition. The inner city, though, is seldom a continuous area nor always central in location. They are generally areas of decline in both population and employment opportunities and have become synonymous with *deprivation*, areas of run-down housing, poor environmental conditions, high levels of unemployment and many other characteristics of poverty and social malaise.

innovation means bringing a new idea into being within either the marketplace or the workplace. For new products the sources of innovation may be based on new technology, new design or a wholly new invention. Process innovation is also of great significance as it can lead to a major cost advantage over competitors. When the British company Pilkington PLC invented a new way of making glass more cheaply and of a far higher quality (the float glass process), it provided not only a direct competitive advantage, but also earned considerable sums in licensing fees from overseas manufacturers.

inputs are the elements that go into producing goods or services, such as *raw materials*, *capital* and the workforce (*labor*).

inselberg: an "island mountain" or an isolated hill standing above an extensive *pediment* or plain resulting from erosion. They are steep-sided and often dome-like in appearance with a marked break of slope at their base. They are found in the humid tropical regions, the savannas, and also desert areas. There is some disagreement about their origin. Some writers argue that they are formed by the parallel retreat of scarp slopes during a long period of erosion; inselbergs would therefore be the residual mass that has not been eroded. Other authorities suggest that they are caused by differential tropical weathering in which the effectiveness of the processes may have been influenced by the density of joint spacing. This weathered regolith would later be removed to expose the unweathered basal rock. This would produce inselbergs as "exhumed" features, brought to the surface by the removal of overlying layers of weathered debris. This latter theory is currently favored.

insolation is the heat energy from the sun – incoming **sol**ar radi**ation**. This consists of the visible spectrum together with ultraviolet and infrared rays. Insolation varies with latitude as a result of variations in the length of the day and the angle of incidence of the sun's rays. These global variations influence the general *atmospheric circulation*. Of the total solar radiation that reaches the outer layer of the atmosphere about 51% reaches the ground surface (34% directly and 17% as indirect or scattered radiation from particles in the atmosphere). About 14% is absorbed by the atmosphere and around 35% is reflected or scattered back by clouds, atmospheric particles and the Earth. Different sources quote different values for these various components. It is important to appreciate that these can only be presented as average figures; latitudinal variations of *albedo* due to the nature of the surface and the density of cloud cover are difficult to determine with precise accuracy.

instability is the tendency of an air mass to continue to rise because it is warmer than its environmental surroundings. When air is forced to rise, for example, over a mountain barrier or along a front, it cools according to *adiabatic* laws. If the rising cools at a slower rate than the environmental air then it will be warmer than its surroundings and will continue to rise, following the principle that warm, or less dense, air rises. When the ELR (*environmental lapse rate*) is high this tends to produce instability. A high ELR is defined as a rate that is greater than the *dry adiabatic lapse rate (DALR)*, i.e. greater than 1°C per 100 meters. Instability leads to greater convectional movement and overturning of the air, which produces deeper cumulus and cumulo-nimbus cloud with heavy rain and more "violent" forms of weather such as hail, thunder and lightning.

intake is an area of upland or moorland that is enclosed or fenced in order to enable some form of land improvement. This could involve drainage schemes to allow more productive use of the land or the use of fertilizers to improve upland pasture.

integrated pest management is a way of attempting to limit pest damage to crops with a much lower use of chemicals. Ways in which pest damage can be limited in such programs include:

- deliberately introducing pest predators into the system
- not using pesticide in a way that also kills the predator
- timing crop planting to avoid the time of the year when the pests are most active
- developing physical methods of destroying pests such as small flame burners or constructing barriers against the pest
- rotating crops to avoid build-up of crop-specific pests
- if *pesticide* has to be applied, then using it at the time when it will have maximum effect, possibly with the aid of computer technology to assess the time of maximum usefulness.

integration: the joining together of different regions or countries in order to gain benefit. The *European Union* is an example of an integrated group of countries, with cooperation taking place over several economic, social, political and environmental policies. The future extent of their integration is the subject of considerable debate within and between each of the constituent countries. Supporters of further integration point to the greater economic and political power that would ensue. Opposition comes from those people who believe that there would be less control over internal affairs of each country.

intensive agriculture is characterized by high inputs of capital and/or labor per unit area. Such levels of input would normally produce high outputs per unit area or high yields. This system may occur with both *subsistence* and *commercial* economies. An intensive subsistence form of agriculture is the production of paddy rice in Southeast Asia. Commercial examples include market gardening, horticulture, dairy farming and battery hen units. Intensive approaches often reflect pressures such as land shortage, high population density or the production of highly perishable foods for an urban market.

interception is the process by which raindrops are prevented from falling directly onto the soil surface by the presence of a layer of vegetation. Water is intercepted by plant leaves, stems and branches particularly in the tree layer but also by species in the shrub or field layer. During prolonged or heavy rain, the capacity of the plant surfaces may be exceeded and water will drip off the leaves and branches (throughfall) and will run along branches and down the trunk (stemflow). *Evaporation* will remove some of the water held on the leaves, a process referred to as interception loss.

interdependence refers to the interrelations within a system. In economic geography, a change or a decision taken in one location can affect the operation of a part of the manufacturing process elsewhere. A fall in price, or overseas competition in the form of cheap imports, can influence the success of an industry. If industry declines then there are knock-on effects on employment, social conditions, health and crime levels. In physical geography a change in one part of the ecosystem brings about a change in other areas. Removing trees may increase run-off into rivers, which changes the pattern of erosion and deposition, altering the system.

interglacial: the period of relatively warmer climate separating two stages of ice advance. Originally three interglacial periods were identified in Europe based on the interpretation of deposits in Alpine valley terraces. These separated the advance phases of the Gunz, Mindel, Riss and Wurm. Investigations of the isotopic ratio between the oxygen isotopes O^{16} and O^{18} in deep sea and polar ice cores has indicated many more, but shorter, interglacials. There may be up to 20 with lengths up to 10,000 years. Conditions in these period have been reconstructed by using pollen analysis on peat deposits preserved in tills.

intermediate technology: the matching of technology in the *Third World* to the needs and skills of the people of the area. Too many projects, developed in the *First World* and applied to the Third, failed because the gap between the people's knowledge and the modern technology was too great. With a lower level of technology it is possible for people to be taught the technical understanding and skills that they require to become self-sufficient. In parts of the *Sahel* region in the north of Africa, successful intermediate technology to combat drought has been the introduction of wind pumps to bring water to the surface.

internationalism: the *global economy* has experienced internationalism as each country's economy has become less self-contained and more part of the global process of change. In an economy that is internationalized there is a strong two-way flow in all trade categories.

International Monetary Fund (IMF): the banker to the world's central banks. If a country needs to borrow money, it can apply for a loan from the IMF. It is then likely that the IMF will send a team of inspectors to the country who will advise on the conditions to be tied to the loan such as cuts in government spending.

interquartile range: a statistical method showing the dispersion of a set of values around the *median* calculated by finding the difference between the upper and lower quartiles. When calculated, 50% of all the values fall within it. The smaller the interquartile range, the more the values are grouped around the median. When divided by 2, the interquartile range gives the quartile deviation.

When data is placed in rank order (highest to lowest):

FORMULA:
$$\text{upper quartile} \left(\frac{n+1}{4}\right) \text{ from top of rank order}$$
$$\text{lower quartile } 3\left(\frac{n+1}{4}\right) \text{ from top of rank order}$$

interstadial: a brief time period during a glacial phase when the climate improves sufficiently for limited glacier or ice sheet retreat. It is shorter than an *interglacial* period but may be long enough for some aspects of vegetation development, as in the Allerod interstadial during the last ice phase in Europe.

intertropical convergence zone (ITCZ): the movement of the *trade winds* toward the equator results in *convergence*. This, combined with intense heating, leads to rising air, which, in turn, is facilitated by *divergence* aloft. This zone of convergence is known as the ITCZ and forms the equatorward side of the *Hadley cell.* (See *atmospheric circulation models.*) The ITCZ is associated with an area of precipitation that moves north and south of the equator through the year following the apparent movement of the overhead sun.

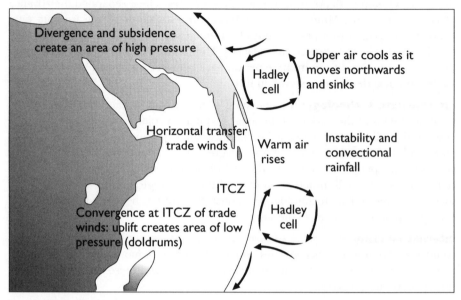

The ITCZ and the Hadley cells

intervening opportunity theory was put forward by Samuel A. Stouffer in 1940 as an attempt to explain levels of movement between places. Interaction between places can be predicted using the *gravity model,* which considers the size of the two settlements and the distance between them as the basis of the calculation. Stouffer suggested that the amount of movement over a given distance is proportional to the number of opportunities at the point of destination, but is inversely proportional to the number of opportunities between the point of origin and the destination. A large number of intervening opportunities reduces the number of movements to the final destination; some of the potential movement is "intercepted" en route.

This theory can be applied to shopping. The presence of closer, good quality shopping outlets will tend to reduce the number of shoppers who are willing to travel to a large, but more distant shopping mall. The formula for predicting movement is:

$$P_{ki} = \frac{O_i}{O_j}$$

where P_{ki} is the number of people living at k who will travel to i to make a purchase, O_i is the number of opportunities to buy the item at i, i.e. the number of stores that sell the item, and O_j is the number of opportunities to buy the item on the route between k and i.

interventionist: an individual who believes that government intervention can help make markets more efficient, and protect individuals from socially irresponsible business behavior. Such a person is likely to promote an active economic policy by government, as opposed to a *laissez-faire* approach.

interventionist policies are those pursued by governments that believe it is their duty to exert a strong influence over the running of a country's economy. Such intervention might include:

- "rescue packages" to help out large firms that have gotten into financial trouble
- prices and incomes policies
- laws to provide stronger protection to consumers or workers
- active support for new firms
- *regional policies*
- tougher laws and higher fines for pollution, etc.
- minimum wage policy

intervention price: a guaranteed minimum price set by a government for an agricultural commodity. Within the *Common Agricultural Policy (CAP)* of the *European Union* the intervention price operates when prices on the open market are falling perhaps as a result of overproduction of a particular crop. When the price falls to the base level of the intervention price, the EU purchases the crop at the agreed minimum price. This is a very controversial policy as it encourages, and rewards at an agreed basic level, farmers who overproduce a crop for which there is no market. This stimulus to production has led to large stores of produce that the media have labeled the beef mountain, the butter mountain and the wine lake. It is difficult to sell off or give away these surpluses to underfed areas of the world because of the detrimental effect this would have on world trade. As it is, producers in other countries object to the level of protection and support the EU gives to its farmers; such policies can destabilize trade and economic stability.

intrazonal: a classification of soils that covers those that reflect the dominance of a single local factor. This includes the influence of geology and extremes of drainage. Typical azonal soils are *rendzina* and *terra rossa*, both developed on rocks that have a high calcium content such as *limestone,* and *gley* soils, developed where conditions are waterlogged.

intrusive: molten rock (*magma*) that is injected into the crust of the Earth from beneath it is said to be intrusive. This may be exposed by later erosion of the overlying rocks. A typical intrusive landform produced in this way is a *dyke.*

inversion: see *temperature inversion.*

invisible export: the sale of a service to an overseas customer. As well as services such as banking, airline travel and insurance, visits by foreign tourists are counted as invisible exports, since they bring in income from overseas.

invisible import: the purchase of a service from an overseas supplier.

invisible trade consists of the import and export of services. They include financial services such as banking and insurance, as well as tourism and shipping. Invisibles account for approximately a quarter of world trade.

inward investment is the *capital* attracted to a region or a country from beyond its boundaries. An example of such investment is a Japanese car manufacturer opening an assembly plant in France.

ionic exchange: see *cation exchange capacity*.

Iron Age: the period that followed the *Bronze Age*, beginning about 500 BC. This cultural period was characterized by the use of iron weapons and tools and was a combination of innovation in indigenous groups together with changes introduced by migrating groups. The most obvious feature of the Iron Age settlement is the hill fort, which represents a focal point of activity, not only in terms of defensive or military functions. The most advanced Iron Age group were the Belgae, who seem to have developed some semiurban functions. Some clearance of woodland occurred during this phase, for iron tools were more effective in clearing larger trees. The Belgae also introduced a heavier plow that allowed some cultivation of the heavier, more fertile soils. Precise dates for this period are difficult to estimate and are regularly reassessed and reinterpreted.

iron pan: a thin layer of mineral grains cemented together by a high concentration of redeposited iron and humus located in the lower horizons of a soil. An extensive iron pan is impervious to water, which causes waterlogging of the soil directly above it. An iron pan is a common feature of the *podsol* type of soil.

irrigation: the artificial watering of the land by humans during dry periods of weather. For many hundreds of years, techniques such as the use of tanks, inundation canals and the Archimedes screw have been slow, primitive and inefficient. Newer developments, involving barrages, large reservoirs, pumping stations and booms, have allowed vast areas of land to be cultivated. Such schemes have not been without their problems, including the silting up of reservoirs and *salinization*. In recent years smaller-scale tube wells have become more widespread in the less developed world.

island arc: a chain of volcanic islands located on the continental side of an *ocean trench*. It is associated with a *subduction* zone where an oceanic plate is being pushed below a continental plate. Subduction causes some of the oceanic plate to remelt, and silica-rich lavas (andesites) rise to the surface and create volcanoes. This situation arises in the western Pacific Ocean where the islands that make up the Aleutians, Japan and the Philippines constitute island arcs.

isodapane: a line connecting points of equal total transportation costs. It is most widely used in an application of the *Weber* model of industrial location when the isodapanes represent the total costs of the transportation of raw materials and the finished product. The critical isodapane in this model delineates the area within which any reduction in costs, for example through economies resulting from cheaper labor costs, must be greater than or equal to the additional transportation costs involved in moving to a place away from the point of least cost.

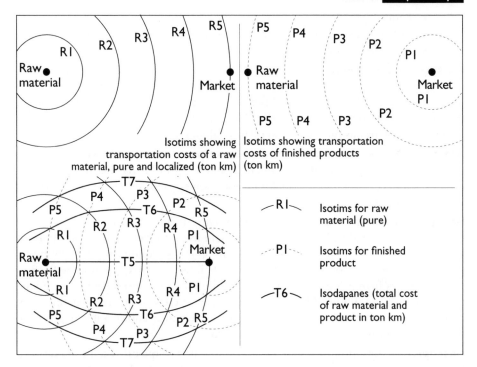

The construction of isodapanes from isotims

isopleth map: chart in which lines have been drawn to show the variations in the values represented. A single isopleth connects the points on the map at which the variable represented has an equal value. Examples of isopleth maps include

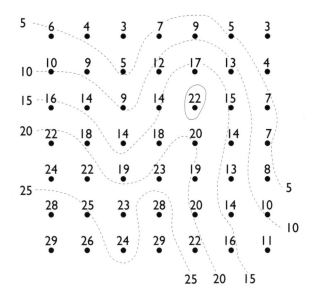

Isopleth map

147

- contour maps connecting points of equal altitude
- isobar maps connecting points of equal atmospheric pressure
- isotherms connecting points of equal temperature.

isostatic readjustment concerns a localized change to sea level. During the major glaciations, the weight of the ice depressed the Earth's crust immediately beneath the ice. This caused a local relative rise in sea level. However, since the ice sheets have melted, the Earth's crust, which had been beneath the ice sheets, has begun to rise. This has caused a local relative fall in sea level.

isotim: a line connecting points of equal transportation costs. As with *isodapanes*, they are generally associated with the *Weber* model of industrial location where an isotim is a line connecting the equal transportation costs for either a raw material or a finished product.

isotope analysis: an examination of the varying forms of an element that have differing atomic weights. In geography, the main use is to identify concentrations of a certain isotope that exist under different climatic conditions. For example, warmer conditions encourage greater accumulations of the oxygen isotope 0–18 in water. If this is frozen into ice in an *ice sheet*, then it will be preserved as a climatic record of such conditions. Other isotopes can be used as a means of dating. The best known is carbon-14, which decays radioactively at a known rate; this can be compared with carbon-12, which does not decay.

isotropic: a description of a state of uniformity. It is frequently used to describe a flat land surface where soil fertility and climate do not vary. There are no physical barriers to movement and so transportation is equally easy in all directions.

IT: see *information technology.*

ITCZ: see *intertropical convergence zone.*

J

jet stream: a band of very strong winds, up to 250 km per hour, occurring in certain locations in the upper atmosphere. A jet stream may be hundreds of kilometers in width, with a vertical thickness of one to two thousand meters. On average they can be found at altitudes of 10,000 meters. They are the product of a large temperature gradient between two air masses that have markedly different temperatures. There are two main locations of jet streams:

- the Polar Front jet – a westerly band associated with the meeting of polar and tropical air above the Atlantic Ocean at approximately latitude 60°N and 60°S. The precise location of this jet stream varies on a daily basis, but airplane pilots may seek to "ride" in it when going from west to east, and to avoid it when flying from east to west
- the Sub-Tropical jet – also generally westerly and associated with the poleward ends of the *Hadley cells* at approximately 25°N and 25°S. However, in summer above West Africa and southern India this jet may become easterly. This is due to temperatures over the land in these areas being higher than over the more southerly sea areas.

JIC: see *"just in case" production*.

JIT: see *"just in time" production*.

Johnson, Douglas Wilson: formulated a simple genetic classification of coasts in 1919, attempting to take account of the origin of coastlines. He divided coasts into four main types:

- shorelines of submergence, where there was a relative rise in sea level due to subsidence of land or a rise in sea level, or both
- shorelines of emergence, where there was a relative fall in sea level
- neutral shorelines – this includes coasts where the exact form is due to neither emergence nor submergence but to a new form of construction or tectonic process: deltas, volcanoes, fault systems
- compound shorelines, including coasts that have an origin combining at least two of the above classes.

In practice, it is difficult to assign coasts using this classification; most of the world's shorelines show signs of both emergence and submergence due to oscillations of sea level during the *Pleistocene* period. Coasts can only be classified according to the most strongly marked characteristic. Most early attempts at classification present difficulties because they are either inflexible or descriptive. *Valentin*, in 1952, proposed a more dynamic approach based upon advancing and retreating coastlines.

joint: a natural vertical crack in a rock. Some rocks are more jointed than others, for example, *carboniferous limestone*. Joints are places that *weathering* and *erosion* processes can take advantage of, and thus may be regarded as lines of weakness in that rock. Well-jointed rocks tend to weather and break up more easily.

"just in case" production: a manufacturing system that involves the stockpiling of raw materials and components at the site of the production process. This was the traditional way of organizing production in many Western countries, and it has proven to be both inefficient and wasteful in many industries. It is being superseded in many cases by the "just in time" production (JIT) system introduced by Japanese industrialists.

"just in time" production: a manufacturing system designed to minimize the costs of holding stocks of raw materials and components by carefully planned scheduling and flow of resources through the production process. It requires a very efficient ordering system and reliability of delivery. Another requirement of a JIT system is that there must be "zero defect" or "total quality."

kame: a *fluvioglacial landform* consisting of a mound of sand and gravel deposited by the *meltwater* from a glacier. The features are similar to a series of *deltas*, and are formed along the front of a stationary or slowly receding glacier or *ice sheet*. At the sides of the glacier, meltwater streams often deposit material that, after the retreat of the ice, form kame terraces. Like all fluvioglacial landforms their deposits show sorting by water.

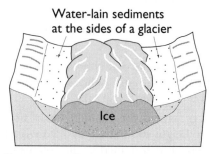
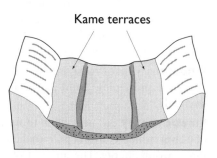

Kame terrace formation

karst: the name given to the scenery that develops on *Carboniferous limestone*. Similar types of *limestones* in different parts of the world produce the same features. The name comes from the Karst (or Carso) region of the former Yugoslavia where this type of scenery is very well developed.

katabatic refers to winds that flow down valley sides and valley floors. This typically happens at night in upland areas as the cool, denser air moves to lower levels. In places like the British Isles this is a very gentle feature, but on glaciers and in such areas as Greenland and Antarctica they can be extremely strong. (See also *anabatic*.)

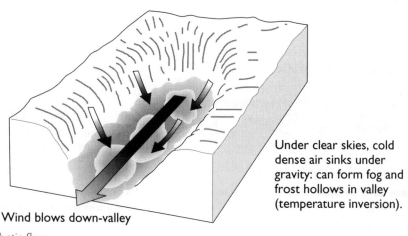

Katabatic flow

kettlehole: a depression in a glaciated lowland area that usually contains a very small lake. Their formation results from blocks of ice that become detached from the icefront as the glacier or *ice sheet* retreats. These blocks become covered in glacial and *fluvioglacial* debris and slowly melt. As they disappear, a hollow is produced on the site, which later fills with water.

Keynesian: a person whose economic ideas can be traced to those of John Maynard Keynes. Oversimplified, these beliefs include:

- government action may be needed to push an economy out of recession
- allowing people to suffer, through unemployment, for example, while waiting for the free market to bring the economy into balance is morally and socially unacceptable
- it is desirable for the government to follow policies that will iron out fluctuations in economic activity. (See *business cycle.*)

kibbutz: a form of rural settlement developed in Israel and organized according to collective principles. Land is commonly owned and farming policies are jointly decided.

kinetic energy or energy due to movement is generated, for example, by the flow of a river, which converts potential energy due to height above sea level or base level into moving energy. The amount of kinetic energy is determined by the volume of flowing water and its average velocity. An increase in *discharge* and/or velocity leads to an increase in kinetic energy.

knickpoint: when a river is *rejuvenated*, adjustment to the new base level starts at the sea and gradually works its way up the river course, the point of change to the existing profile being known as the knickpoint. Rejuvenation occurs when the *base level* falls as a result of a fall in sea level, or the uplift of the land. Either way, the river gains renewed cutting power, which encourages it to adjust the *long profile*. In that sense the knickpoint is the place where the old long profile joins the new. This is usually marked by rapids or a waterfall. If the process has happened more than once, there could be several knickpoints on a river.

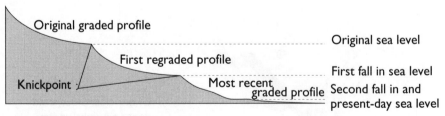

Knickpoints on the long profile of a river

knock and lochan: a landscape made up of small rocky hills (knocks) and small lakes (lochan). It is the result of erosion by an *ice sheet* on a surface where there are slight variations in the strength of the rock.

kolkhoz: in the former *USSR*, the name for a *collective farm*. In the 1920s, these farms replaced individual farmers and the large estates, as the state ownership of land was fundamental to communist ideology. The land was leased from the government and managed by a committee led by a chairperson. As many as 400 families could be working on one farm. The produce was sold to the state at a fixed price, with families sharing the resulting income. They were also able to grow their own food (or small amounts for sale) on small plots on the kolkhoz. Most of these farms were found toward the west of the USSR, particularly in Ukraine. They often failed to meet production targets set by the state and from the 1940s were increasingly replaced by state farms or *sovkhoz*.

Kondratieff cycle: the theory that in addition to the five-year *business* (or trade) *cycle*, there exists a 50-year cycle of economic upturn and downturn. This theory was put forward by the Russian economist Nikolai Kondratieff in the early 20th century. It was dismissed by many economists until the Great Depressions of the 1880s and 1930s were duly followed (50 years on) by the frequent and severe recessions of the period 1975–1992. The most widely accepted explanation for the cycle is that the introduction of new technology causes disruption, but once established it forms the basis for many new products and jobs. In the 1930s the car was displacing rail, while in the 1980s the microchip was replacing mechanical technology.

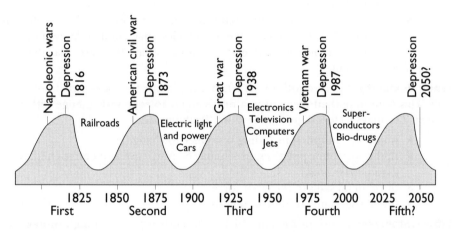

Kondratieff cycles

kurtosis is a measure of the extent to which the general shape of a *frequency distribution* graph is peaked. A distribution that has a relatively narrow but high peak is termed leptokurtic; one that is basically broad and flat is termed platykurtic; one that is between these two extremes is called mesokurtic. Distributions can also be described according to their degree of symmetry, or positive and negative *skew*.

L

labor is one of the *inputs* that goes into producing goods and services. There are various aspects of labor, many of which vary spatially and are therefore important elements in the *location of industry*. Some of these elements are:

- cost – the cost of labor to a company can be between 15 and 40% of its total costs. Differentials in global wage levels can be an influential part of the decisions taken by *transnational companies* in locating plants
- quality – access to a skilled workforce has always been an important factor for some industries. Today, it is a very important consideration for many small-scale *light industries*, particularly those involved with computers and *robotics*
- relations – centers of union activity are regarded by some employers as potential centers of militancy and therefore as a negative location factor. This has tended to become less important with the decline in labor union membership
- availability – some areas may have the required workforce for a potential employer. For example, some companies may want a large number of female employees and go to regions where they are available for employment.

labor flexibility: the ease with which a firm can change the jobs carried out by its staff. This is an important element in a firm's ability to cope with change within its market. The main factors determining labor flexibility are:

- workforce attitudes, including resistance to change
- traditional labor practices, which may be restrictive and entrenched, such as job *demarcation*
- the general skills of the labor force
- the strength and attitude of unions.

labor-intensive: a work process in which labor costs represent a high proportion of total costs. Such a situation is most likely to exist in the service sector. This contrasts with firms that are *capital*-intensive.

labor market is the supply of labor (i.e. all those offering themselves for work) and the demand for labor (i.e. employers), which together determine wage rates.

labor mobility is the extent to which labor moves around in search of employment, called geographical mobility, or the extent to which labor moves between jobs, which is known as occupational mobility.

Geographical mobility depends on such factors as availability of housing, costs of moving, family ties, and the availability of information, while occupational mobility depends on training facilities and a willingness to learn and be retrained.

lag time is the period between the maximum precipitation within a *drainage basin* and the peak discharge of the river in that basin. The information is displayed on a *hydrograph*. The lag time varies according to conditions within the drainage basin such as soil and rock type, slope gradient, drainage density, type and amount of vegetation, extent of urbanization, size of basin and water already in storage, which will in part depend on recent weather.

lahar: a flow of wet material down the side of a volcano's ash cone. Lahars occur when surface water picks up large amounts of volcanic ash and deposits it as mud over lower-lying areas. In the aftermath of the eruption of Mount Pinatubo (Philippines) in 1991, *typhoons* brought heavy rainfall that caused a number of lahars to form.

laissez-faire is a political and economic philosophy that asserts that governments should avoid interfering in the running of business or any other part of the economy. A laissez-faire economist places faith in the ability of the free market to maximize business efficiency and customer satisfaction. This is in direct contrast to the views of *interventionists*.

lame duck: a firm or industry that cannot compete and is therefore likely to go into liquidation without the help of the government. Some governments have argued that it is the market that should ultimately decide which companies should survive and which should not. The advantages and disadvantages of supporting lame ducks are:

Pros: • industries or firms that qualify for help are often large employers and financial support may avoid the long-term financial and social costs associated with unemployment
• if large firms fail they may take other firms with them that could have continued to be efficient and profitable
• firms often go through difficult periods, which does not mean that they are inefficient. Helping such firms does not mean a long-term financial commitment.

Cons: • the resources given to struggling firms could be used more effectively in growth industries
• if firms know that they can fall back on governments, they may lack the incentive to become efficient and profitable
• it is often only large firms that receive such help
• experience shows that civil servants do not have a good record in distinguishing between a lame duck and a dead one!

land and sea breeze: a diurnal circulation of air resulting from the differential heating and cooling between land and adjacent sea areas. By day, the land heats up more quickly than the sea. Hence the air over the land becomes warmer than the air over the sea, and low pressure forms over the land, with relatively high pressure over the sea. A wind therefore blows from the high pressure over the sea to the low pressure over the land. This is a sea breeze, which can lower the temperatures of coastal areas by several degrees.

By night, the land cools down more quickly than the sea. Hence the air over the land becomes cooler than the air over the sea, and relatively high pressure forms over the land, with low pressure over the sea. A wind therefore blows from the high pressure over the land to the low pressure over the sea. This is a land breeze.

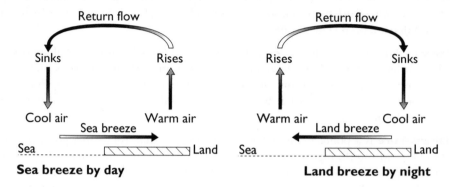

Land/sea breezes

landfill site: a land-based location used for the disposal of waste. The purpose is to empty waste materials into a natural or man-made hole in the ground, for example a disused quarry, and to cover it with soil when full. However, hazardous materials may infiltrate the soil and enter *groundwater* supplies. In addition, the decomposition of the waste under the ground in anaerobic conditions causes methane to be produced. This rises to the surface and presents a potentially explosive hazard.

land reclamation includes any process that improves or recovers land usually for agricultural use, although *derelict land* in urban areas may be reclaimed for industrial, recreational or residential purposes. Drainage may be installed to improve low lying or marshy land, infertile areas can be treated with chemicals, sea areas and lakes can be drained to produce new agricultural land, and moorland areas can be ditched to aid surface run-off.

land reform: a term used to describe the redistribution of land in an area in order to increase agricultural productivity and to raise individual standards of living. In many countries earlier this century there was an unequal distribution of land. Ownership was concentrated in a small number of large estates, while the bulk of the rural population had too little land for a reasonable living. Thus a number of wholesale changes in the ownership of land have taken place. In the communist countries the landlord had his land expropriated, and redistributed among the peasant farmers. In some of these, the smaller peasant farms were then grouped together into collectives. In other countries the government has compulsorily purchased private land, and resold it under favorable credit terms. An increase in owner-occupancy is perceived as being an incentive to raise agricultural productivity.

LANDSAT is an American satellite orbiting the Earth at a height of 705 km. It is an unmanned remote sensing device that can monitor features on the surface.

landslide: a type of mass movement in which the debris moves downslope as a unit along a glide plane. They occur particularly where the rocks have *bedding planes* that are roughly parallel to the slope angle. They are activated by the accumulation of large amounts of soil water in the weathered surface material. This added weight increases the stress on the debris and the extra water reduces friction between the particles and lubricates the mass of material along the bedding plane.

landslip: an alternative term for *landslide*.

land use: any purpose for which land is used for human activity. Such uses can be divided into rural – agriculture, forestry, settlement, recreational space – or urban – housing, industry, commercial.

lapse rate: the rate at which temperature decreases with height. In the atmosphere the change of temperature with height is called the *environmental lapse rate (ELR)*. Air that is moving through the atmosphere will cool on rising, or warm up on descending, according to adiabatic laws. When the air is dry, i.e. it has not become saturated by cooling to its *dew point*, it changes at the *dry adiabatic lapse rate (DALR)*. Once it is saturated it will cool or warm according to the *saturated adiabatic lapse rate (SALR)*.

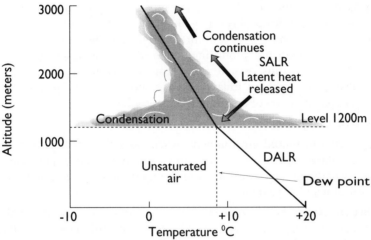

Lapse rates

latent heat is the heat energy expended when a substance changes its physical state without raising its temperature. During the process of evaporation when liquid water or ice changes to water vapor, heat is required to bring about the physical change from water to vapor. During condensation, when the process is reversed and water vapor condenses into water droplets, latent heat is released. Latent heat of evaporation causes cooling while condensation produces a heating in the atmosphere. This large-scale release of heat can accelerate rising air currents in *cyclones* and *convection* systems and influences the *saturated adiabatic lapse rates (SALR)*.

laterite is a hardened layer of ironstone formation in soils of tropical regions. It has resulted from the precipitation of iron and aluminum minerals either at the surface or in the subsoil. This may produce a *duricrust* layer. A lateritic horizon may be produced where the groundwater movements in the soil concentrate iron and aluminum oxides into a depositional zone that coincides with the depth of the *water table*. These concentrations may appear as nodules or as a slag-like layer. The term is also used in a general way to distinguish tropical soils, such as latosols, which are the result of the process of laterization.

latifundia: a system of land holding that consists of large estates on which a large number of the local people of an area work. The system was common in the south of Italy until recent times, and was often cited as one of the reasons why the region was so agriculturally backward. The *Cassa per il Mezzogiorno* was given the responsibility for the dismantling of the system in order to promote greater farm efficiency and the emergence of a new landowner class. The system still flourishes in parts of Latin America, particularly Brazil.

Laurasia is the name given to the northern supercontinent that consisted of present-day North America, Greenland, Europe and most of Northern and Central Asia. As *Pangaea* divided, Laurasia was separated from *Gondwanaland* by the Sea of Tethys.

lava: when molten rock (*magma*) flows onto the surface it is known as lava. There are several forms, including:

- acid lava, which quickly solidifies on contact with the air and produces steep-sided volcanic forms (e.g. the spine of Mt. Pelée)
- basic lava, which is basaltic in character and tends to flow a long way before solidifying, producing rather flat volcanic shapes such as the Hawaiian Islands, as well as extensive lava plateaus such as the Antrim Plateau in Northern Ireland (the basaltic flow can be seen at the famous tourist attraction of the Giant's Causeway).

LDC: see *least developed country*.

leaching occurs in humid environments when rainfall is greater than *evapotranspiration* and soluble bases are removed from a soil by downward percolating water. The water is naturally slightly acidic, with the most common bases removed being sodium, calcium and potassium.

least cost location: the centerpiece of Alfred *Weber*'s model of industrial location since he stated that an industrialist would seek to establish a factory at that point. It is the point at which production costs – the combination of raw material and transport costs – are at a minimum.

Weber also stated that an industry may be located away from the least cost location when:

- cheaper labor costs exist at another point, which result in greater savings than the additional transport costs to that point
- cost savings from *agglomeration* also outweigh the additional transport costs involved.

least developed country (LDC): a concept first identified in 1968 by the United Nations Conference on Trade and Development, but subsequently updated by other United Nations international conferences. The main characteristics of such a country are:

- a low *GDP* per capita ($100 or less)
- a low level of *literacy* (less than 20% of the population aged 15 or over)
- a low percentage of manufacturing in the economy (less than 10%)
- a low *life expectancy*
- a high *infant mortality* rate

- a high rate of annual population growth resulting from a high birth rate and a falling death rate
- a low provision of medical facilities.

Several least developed countries are also physically isolated from the rest of the world, for example by being landlocked such as the *Sahel* countries of Africa. Other features are that most suffer from trade instability, and several have political unrest. (See *economically less developed country.*)

leisure industry: the employment and products generated by the activities people engage in for enjoyment during time that is regarded as being free from work demands or any other obligation. The leisure industry has expanded considerably in recent years in economically more developed countries as people have both more spare time and more disposable income. An increase in personal mobility has also helped this to occur. The leisure industry is more than just tourism. People engage in "at-home" activities, as well as activities outside of the home. Some pursuits are active, whereas others are passive. Since the industry has grown at such a rapid rate, concern is now being expressed at its impact on the environment. (See *honeypot* sites.)

less favored areas are regions within the *European Union* that are given extra funding because they are difficult environments to live and work in.

lessivage: a particular kind of *leaching* that results from clay particles being carried downward in suspension. This process can lead to the breakdown of the *peds* (the aggregation of different particles that give the soil its structure).

levée: that part of the bank of a river that is raised higher than the *flood plain*. They form naturally when the river floods as the sediment is dropped. It is usual for the coarsest to be dropped first, forming a small bank along the channel. Subsequent floods increase the size of this bank. The river, with channel sediment build-up, will now flow at a higher level than the flood plain; therefore authorities sometimes strengthen the levée and also increase its height. On the Mississippi River, for example, the levée strengthening began in 1699. By 1738 they had built 68 km of bank. In the 1990s the length of the levées is now 3200 km.

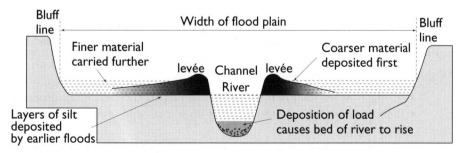

Position of levées on river flood plain

level playing field: a phrase that sums up the need within a *market economy* for all firms to be competing on the same terms. If one national government is subsidizing its steel producers, for example, they will have an unfair advantage over those from a country that offers no state aid to the steel industry. The *single European market* is an attempt to provide a level playing field in relation to laws and regulations governing traded goods within the *EU.*

lichen: a very simple plant that will be one of the *pioneer* species when bare rock areas (*lithoseres*) are colonized. Lichens are very undemanding plants, as they are capable of living without soil, with no supply of permanent water and in extremes of temperature. Lichens help in the break-up of the rock surfaces. As conditions improve with the creation of a thin layer of soil, other plant species begin to take over from the pioneers as a *succession* develops, although lichens may be part of that new community.

life expectancy is the average number of years from birth that a person can expect to live.

light industry is that part of *manufacturing industry* that does not fall into the classification of "heavy." *Heavy industry* is one in which large weights of materials are handled; therefore light industry handles small amounts of material per worker involved. Light industries are not generally sources of pollution and are often found in modern locations such as industrial parks.

limestone: a *sedimentary rock* composed wholly or largely of calcium carbonate and formed by:

- the accumulation of the skeletal remains of marine creatures, for example chalk
- chemical precipitation, for example oolitic limestone.

Both types are usually accumulated in layers and compacted. They may also contain other materials such as clays and quartz. (See also *carboniferous limestone, chalk, coral reef.*)

linear: a description of a feature that has long and narrow dimensions. A linear settlement is one that is strung out along a routeway such as a road or waterway, or along a confined river valley. Settlements that seek to avoid flooding also extend themselves along a raised terrace. Many towns have grown outward in a linear pattern alongside a major road. This is known as ribbon development.

linkages refer to the relationships between one industry and another. Linkages may take a variety of forms:

- they may be to the consumer of the industry's product (forward linkages)
- they may be to the provider of raw materials and components (backward linkages).

Linkages can also be classified as:

- vertical – when the raw material goes through several successive processes
- horizontal – when an industry relies on several other industries to provide its component parts
- diagonal – when an industry makes a component that can be used in several subsequent industries
- technological – when a product from one industry is used as a raw material by a number of subsequent industries that further reprocess it.

Linkages may also involve the subcontracting of work, maintenance links, financial links or the use of common services such as packaging and wholesaling.

literacy is the ability of a person to read and write. The literacy rate of a country is used as a means of measuring its level of economic development. It is an indication of the quality of education in a country, which in turn is an indicator of the wealth and social development of that country. Literacy may vary between different social groups, and between different sexes in a population.

lithology: the study of rocks related to their physical, chemical and textural character. It may also be used to describe a rock in terms of these features. A rock's lithology can be described by its permeability, solubility and relative hardness. Its lithology is a major factor in determining the nature and topography of the landscape it produces.

lithosere: a primary plant *succession* that takes place on a newly exposed rock surface. The surface may be rock exposed by a retreating glacier, or on lava following a volcanic eruption, or on new land emerging from the sea. An example of a lithosere in a temperate part of the world would be as follows:

- the area of bare rock is colonized initially by bacteria that can survive where there are few nutrients
- lichens and mosses also colonize the rock, and assist in the weathering of the rock to form a thin layer of soil in which more advanced forms of plants can grow (see also *pioneer* community)
- as these plants die, the bacteria convert their remains into humus, which results in a more fertile soil
- grasses, herbs and small flowering plants then colonize the area, which in turn give way to shrubs
- these shrubs are replaced by faster growing trees such as rowan, which then face competition from slower growing trees such as ash and oak. These trees eventually form the climatic *climax vegetation* for the area.

lithosphere: a layer of rigid rock that forms the solid upper part of the Earth's *mantle* and *crust*. It is about 100 km deep, and "floats" on top of the semimolten *asthenosphere*.

livestock are animals kept or farmed for use or profit. Livestock farming may be extensive, as in ranching on the Pampas grasslands of Argentina, or it may be intensive, as in dairying in northwest Europe. Livestock may also be kept by *nomadic* herders who generally lead a subsistence existence. Some livestock may also be kept as draft animals.

load: the material transported by a river. It may be transported by:

- *suspended load*
- *saltation*
- *solution.*

localized: a term applied by Alfred *Weber*, in his theory of industrial location, to a raw material that is found at particular points in the landscape. Such raw materials exert a strong influence on the location of industry because of the transport costs incurred in moving them to the point of manufacture. Examples include coal and iron ore, the basis of traditional heavy industries. Other materials that are found everywhere on the landscape are termed *ubiquitous.*

location quotient: a statistical measure of the geographical concentration of an activity in a region compared with the national average. The LQ is calculated using the formula:

$$LQ = \frac{X_A/Y_A}{X/Y}$$

If the LQ was applied to the concentration of industrial activity:

- X_A = employment in industry X in region A
- Y_A = total employment in all industries in region A
- X = national employment in industry X
- Y = total employment in all industries nationally.

This can be simplified as

$$LQ = \frac{\% \text{ of total workforce in region A working in industry X}}{\% \text{ of total national workforce working in industry X}}$$

If textiles employs 12% of the workforce in the region, but nationally 4% of the total workforce are employed in this industry, then LQ = 12/4 = 3.

An LQ of 1 indicates that an industry is represented in a region in exactly the same proportion as nationally.

Less than 1 suggests that the industry is underrepresented; the region has less than its fair share of the activity.

Greater than 1 suggests that the industry is overrepresented; there is a geographical concentration and the region has more than its fair share.

LQs are useful in measuring the relative concentration of an activity at a particular time, but care must be taken when comparing different time periods. Because there are four components to the equation an industry in a region could increase its workforce but produce a lower LQ if its rate of increase in the industry was lower than the national rate of change. Interpretation only relates to the degree of concentration.

lodgment till is a form of ground *moraine* deposited under a glacier or ice sheet. Toward the snout of a glacier, as the forward velocity is reduced bottom melting releases rock particles from the sediment-enriched basal layer of the ice. These are laid down under pressure beneath the slowly moving ice, plastered to the bedrock. Because of the slow movement of the ice above the lodgment layer, the till stones may display some orientation with their long axis in the direction of ice movement. This contrasts with the more haphazard deposition of *ablation* till.

loess is a wind-blown (aeolian) deposit. It is regarded as a *Pleistocene* sediment derived from the outwash plains of the last ice advance. It stretches across Europe from the Paris Basin, where it is called limon, through Belgium and south Germany into Poland. It is a yellowish brown loam, which is easily weathered to produce very fertile soils. In China it has produced massive deposits up to 300 meters thick. Loess is a relict deposit; this form of wind action now seems to be less important and loess is not being added to under present-day conditions.

logarithmic scale: a scale divided into a number of cycles, each representing a tenfold increase in the range of values. If the first cycle ranges from 1 to 10, the second cycle extends from 10 to 100, the third from 100 to 1000 and so on. For very small values, for example when plotting soil grain size analysis, the cycles could extend from 0.001 to 0.01, 0.01 to 0.1, 0.1 to 1.0. Zero cannot be plotted on logarithmic graph paper, nor can positive and negative values be shown on the same axis. Logarithmic scales are useful when the rate of change is of more interest than absolute change; a steeper line reveals a faster rate of change. They also allow a wider range of data to be displayed than on a similar sized piece of arithmetic graph paper. If the rate of change being displayed is increasing at a constant proportional rate, i.e. the population doubles in each unit of time, this will appear as a straight line if plotted on semi log paper (one axis arithmetic for time, the other logarithmic for population). When the perfect *rank size rule* relationship is plotted on semi log paper it produces a straight line with a gradient of −1 – that is, the line is descending at an angle of 45° to the horizontal.

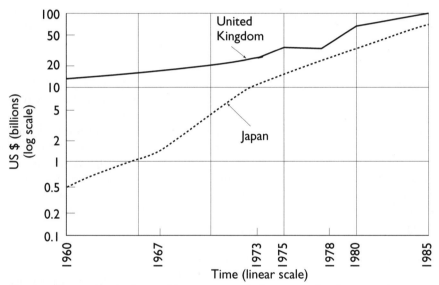

Overseas investment in the world economy made by transnational corporations based in the United Kingdom and Japan

Example of a logarithmic graph (using semilogarithmic paper)

longevity is the increase in life expectancy. With increased economic development and improved medical provision there have been significant changes in mortality in recent years. Life expectancy at birth in the USA in 1950 was 66.0 (males) and 71.7 (females). By 1995 this had risen to 72.8 (males) and 79.7 (females). With declining or steady birth rates this greater life expectancy produces an aging population with a greater percentage of the population over 60. This is sometimes referred to as the "demographic time bomb": there is a smaller percentage of economically active population to support an increasing retired and elderly population. As increasing numbers are living beyond the age of 75, an age when many become more dependent on the state or family help, the cost of health provision and care is high. Many retired people do have private incomes through pensions, creating a group of consumers with considerable purchasing power and influence, the so-called "gray market."

long profile illustrates the changes in the altitude of a river's course from its source along its channel to its mouth. In general a long profile is smoothly concave in shape, with the gradient being steeper in the upper course and becoming progressively less steep toward the mouth. However, irregularities frequently exist, represented by waterfalls, rapids, or lakes. There may also be marked changes in slope, known as *knickpoints*, which are a product of *rejuvenation*. (See *graded profile*.)

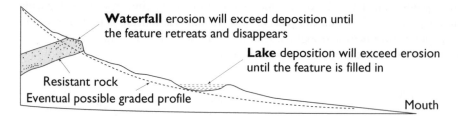

A typical long profile of a river

longshore drift is the movement of sediment along the coast by wave action. When waves approach the shore at an angle, material is pushed up the beach by the *swash* of the breaking wave in the same direction as the wave approach. As the water returns down the beach the *backwash* drags material more directly down the steepest gradient, which is generally at right angles to the beach line. Over a period of time sediment moves in this zigzag fashion down the coast. Obstacles such as *groins* and piers interfere with this drift and accumulation of sediment occurs on the windward side, leading to entrapment of beach material. Deposition of this debris takes place in sheltered locations, such as at the head of a bay, and where the coastline changes direction abruptly when spits develop.

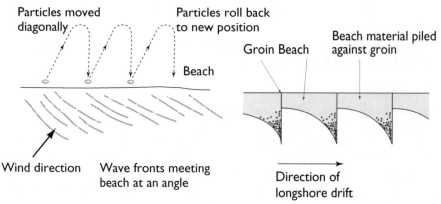

Longshore drift

long-term climatic change takes place over several hundreds, thousands or even millions of years. Evidence for such change over a long timescale comes from the presence of coal and limestone that must have formed in environments much different from the present. The climatic change that has taken place since the end of the Pleistocene *Ice Ages* is also an example of long-term climatic change. Several theories exist to explain these climatic changes. Some state that there have been variations in the solar energy emitted; others state that there have been cycles of celestial movements of the sun and the Earth. However, no clear consensus of opinion exists.

lorenz curve illustrates the degree of unevenness in a geographical distribution. It is drawn on graph paper and makes use of cumulative percentage data. The vertical axis carries the cumulative data, and points are plotted in the order of the largest, which is then added to the second largest, then to the third largest, and so on. The horizontal axis simply records the cumulative process. The plots are then connected by a line. If another line is drawn onto the graph to represent an even distribution, then the degree of unevenness can be seen. The greater the deviation the plotted line has from the line of even distribution, the greater the degree of unevenness. A highly concave lorenz curve represents a high level of unevenness, and therefore high level of concentration.

Assume a country has 10 regions A to J:

REGION	A	B	C	D	E	F	G	H	I	J
% of national population in region	10.5	1.6	12.2	1.8	35.3	6.7	2.1	25.3	3.5	1.0

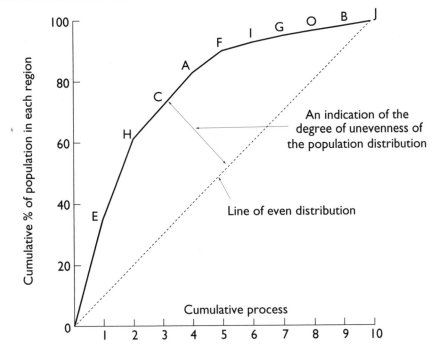

Lorenz curve – an example to show the unevenness of the population distribution of a country

Losch, August: an economist who modified the earlier work of Walter *Christaller* on *central places*. Like Christaller, he used hexagonal market areas but allowed various hexagonal systems to coexist. Each system operates at a different level and is superimposed on the other. Losch's more variable and complex system produces a continuum of settlement sizes that more closely relates to reality, rather than a stepped distribution as in Christaller's model.

Maastricht is the name of the Dutch town where the treaty on *European Union* was signed by all the member states of the EU in December 1991. The Treaty envisaged the introduction of a single currency by 1999 and formulated a number of areas for combined policy action including industrial and social policy, health and education. The Maastricht Treaty attempted to push the pace of European unity forward, but perhaps too quickly for countries such as Denmark and Britain. In addition, the events of September 1992, when the British pound was forced out of the Exchange Rate Mechanism (ERM), suggested that monetary union was still a long way off. The Treaty came into force on November 1, 1993.

magma is the name given to molten material under the surface of the Earth that has risen from the *mantle*. It is the source of *igneous* rocks.

malaria is an infection that occurs in people who have been exposed to any of several species of protozoan parasite transmitted by the female anopheles mosquito. Malaria is an acute chronic infection that, although preventable, may be fatal if left untreated or treated too late. If treatment begins early it can be cured, but if treatment is inadequate or inappropriate, then the infection may recur over a number of years. In many parts of the world the disease is endemic, i.e. common, and in effect part of the environment in which people live. Sometimes malaria spreads through a community as an *epidemic*, where the number of deaths from it may rise substantially, particularly among the very young. In adults, the frequency of attack may lower resistance to other diseases, and it is these and not malaria that are fatal.

It is often assumed that malaria is a disease exclusively of the tropics, but there is evidence to show that it had a much wider occurrence in the past. The disease will not occur in colder conditions, though, as the parasite is not able to develop inside the mosquito. The management of the disease can be seen in terms of:

- avoidance and prevention – preventing the spread of the mosquito by spraying aircraft, for example. In malarial areas, people can try to prevent being bitten by using mosquito nets at night or suitable clothing. Drugs such as chloroquinine can be taken by people before traveling to countries where malaria is endemic
- treatment – quinine is probably the most effective, once the parasite has entered the body
- control and eradication – controlling the mosquito means attacking the breeding areas with insecticide (DDT was perhaps the best-known, but there is evidence that the mosquito has developed some resistance) or draining ponds and marshes and removing vegetation where the insect might shelter or breed from the vicinity of rivers and marshes.

mall: also called a shopping mall or shopping center, this is a planned development, often under one roof, that contains *retail* outlets along with other services such as banking, restaurants, movie theaters and other entertainment facilities, as well as ample parking. The first mall, the Country Club Plaza, opened in 1922 near Kansas City, Missouri. The largest mall in the USA, with over 400 stores, is the Mall of America in Bloomington, Minnesota, which opened in 1992.

Malls are usually sited well away from the *central business district (CBD)* of a town or city, and have been blamed for the economic decline of many CBDs.

malnutrition results from some form of diet deficiency because either the quantity of food intake, as measured in calories per day, is too low, or there is insufficient balance between proteins, energy foods, vitamins and minerals. Although proteins are needed for building body tissue, the most vital element in the diet is carbohydrate, which provides energy. If this falls below minimum needs body functions divert some of the protein intake to make up for the deficiency. When the intake of protein and carbohydrate are satisfactory the minimum requirements of protein and minerals are normally achieved. Simple measures of calorific intake per capita per day do not present the full picture; the proportion of food from starch and proteins is also important. Malnutrition weakens resistance and exposes people to the dangers of killer diseases.

Malthus, Thomas: the author in 1798 of *An Essay on the Principle of Population as It Affects the Future Improvement of Society,* which was an attempt to show the link between population and *resources.* Malthus based his theory on two principles:

- population, if unchecked, grows at a geometric or exponential rate,
  i.e. $1 \rightarrow 2 \rightarrow 4 \rightarrow 8 \rightarrow 16 \rightarrow 32$
- food supply increases at best at an arithmetic rate,
  i.e. $1 \rightarrow 2 \rightarrow 3 \rightarrow 4 \rightarrow 5 \rightarrow 6$.

When the population growth outstrips the capacity of the resources, Malthus suggested that the growth will be curbed by preventative checks, limiting population growth (postponement of marriage, moral restraint in terms of sex, vice, contraception) and positive checks, which will reduce the population size, such as famine, disease and war. The predictions of Malthus did not come true in the 19th century

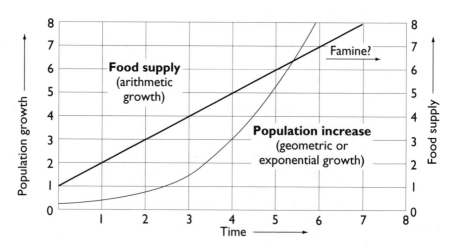

Relationships between population growth and food supply (after Malthus)

for a number of reasons, including the vast improvements made in agriculture, the opening up of new agricultural lands in North America in particular, and the *emigration* of millions from Europe to parts of the world that were considerably less densely populated. In the late 20th century, however, there has been a revival of his views under the heading of Neo-Malthusianism. This was as a response to the dramatic acceleration in population growth in *Third World* countries after their mortality rates began to drop. Rapid population growth was seen to impede development and to bring about a number of social and economic problems. This led to countries in those areas taking positive steps to reduce population growth, assisted by a number of international agencies.

mangroves are coastal wetland forests made up of salt-adapted evergreen trees found in the intertidal zones of tropical and subtropical latitudes. The trees have interlacing stilt or aerial roots that trap sediments and sometimes lead to the creation of new land on the shoreline. They also play an important role in protecting that shoreline from erosion. Mangrove swamps support much wildlife in the nutrient-rich waters and are also a supply of wood for building and for fuel to the local inhabitants. Mangroves have been under threat for a number of years and for a variety of reasons:

- the clearance of the land for agriculture, particularly rice cultivation
- excessive demand for *fuelwood* from increasing populations
- conversion of the swamp areas to fish ponds (*aquaculture*)
- commercial exploitation, particularly by the Japanese, for timber to produce wood chips for cellulose and paper-making.

mantle: the layer of the Earth between the crust and the core. The mantle is separated from the crust by the *Mohorovicic* discontinuity and consists of two layers, the upper or *lithosphere* (together with the crust) and the lower or *asthenosphere*, which is in a semimolten state. The mantle extends to a depth of 2900 km. It is composed mainly of silicate rocks, rich in iron and magnesium.

manual worker: an employee who works with his or her hands, usually in a factory context. Manual workers are often divided into those that are skilled (electrician, welder), semiskilled (van driver, production line worker) and unskilled (cleaner, garbage collector).

manufacturing industry consists of companies that convert raw materials into finished goods or that assemble components made by other manufacturing companies. Manufacturing represents the *secondary sector* of employment.

maquis: a type of vegetation growing in the European *Mediterranean* areas. It grows where there is *impermeable* rock, such as granite, as opposed to *garrigue*, which is found on more permeable rocks. Maquis tends therefore to be taller and denser. Like garrigue, this is said to represent a *plagioclimax* because of man's interference with the natural vegetation.

marina: a dock or basin providing a mooring for pleasure boats. With the increase in leisure time seen in the late 20th century, there has been a great deal of pressure for the development of such facilities. Many urban waterfronts that have deteriorated and declined in the 20th century have been revitalized and often include a marina as part of that redevelopment.

marine (maritime) climate: the climate of those areas lying next to the sea that come under its influence. In *temperate* areas of the world this means a cooler summer than the interior parts of the continent and a milder winter. Coastal regions are often wetter than those further inland, particularly where the coast is backed by a mountain range, such as in the Pacific Northwest. (See also *continental climate*.)

market: the demand for a good or a service. The term can also refer to a specific place that is the market for a good or service. Specific car assembly plants, for example, can be the market for one manufacturer of car components.

market economy is an economy that allows *markets* to determine the allocation of resources. The principal benefits of the market mechanism are that it is automatic and leads to greater efficiency. Markets, however, are not wholly automatic and are always modified or regulated in some way by governments. The extent to which market regulation takes place is a political decision. In the countries that were run as *command economies* (the former communist states), free markets were almost completely superseded by central planning. This system proved unsuccessful, however, not only because it was expensive and inefficient to operate, but also because it stifled individual initiative, and ultimately led to lower living standards. A compromise between a wholly planned and a market economy is called a *mixed economy*.

market processes operate in an environment where the ability to pay the going rate will take precedence over any local or national concerns. Objectors to a planning proposal cannot afford to outbid the developer. Consequently the latter very often succeeds and the development goes ahead with the minimum of consultation. Consultation under market processes often takes the form of an opportunity to voice objections or counterproposals, but with no right of arbitration or right of appeal.

marram grass is associated with the *stabilization* of coastal sand *dunes*. This species is able to tolerate a dry, mobile habitat of shifting sands. Once it roots its horizontal stems or rhizomes through the sand, it binds the sand together, allowing further colonization by other species of grass.

Marxism is a view of society and economic development put forward by Karl Marx in the mid-19th century. He regarded the economic means of production as the key to understanding society and its class structure and patterns of behavior. The population was seen as a commodity or resource in that it provides labor, an important factor of production. As with any other resource, labor costs money to produce in that children require feeding, clothing and educating, and workers need housing. However, in return for this expenditure the capitalists are able to use the resource, and Marx claimed that labor is the only resource that produces a surplus, i.e. the value of the resource to the user is greater than its production cost. As labor is such a "unique" resource, population growth could be viewed as beneficial to economic growth. Marx was not greatly concerned by population growth in relation to resources, unlike *Malthus*, because population provides a means of exploiting resources. Marx considered resources to be sufficient for population, but resources were unevenly distributed.

mass production: the system devised by Henry Ford that turned raw materials into finished products in a continuous moving and highly mechanized process. Its greatest advantage is that it has a high productivity, but this relies on manufacturing very large numbers of an identical product for a mass market. Another feature is that workers on the assembly line are given single tasks for the sake of production efficiency. One of the most famous phrases of Henry Ford epitomizes the spirit of mass production, when he said of his Model T car, "You can have any color you want ... so long as it is black." In recent years, mass production has given way to more *flexible manufacturing systems*. (See also *Fordism*.)

mass wasting (mass movement) is the downslope movement of weathered material under the influence of gravity. One way of classifying such movement is by its speed and nature:

- slow flows such as *creep*
- rapid flows such as *earthflows* and *mudflows*
- *slumping*, *landslips* and *landslides*
- *subsidence* and *avalanches*.

The rate of mass wasting depends on the degree of cohesion of the weathered material, the steepness of the slope down which the movement takes place, and the amount of water contained in the material. A large amount of water adds weight to the mass, but more importantly lubricates the plane along which movement can take place.

material index (MI) was devised by Alfred *Weber* in his theory of industrial location. It is calculated by dividing the weight of the *localized* raw materials needed for one unit of output by the weight of that unit of output (i.e. the product). Where the MI is greater than one, the industry is located at the raw material. Where the MI is less than one, the industry is located at the market. Where the MI equals one, the industry is located at an intermediate location, or at the market, or at the raw material.

mean: see *arithmetic average*.

meander: a sinuous bend in a river. An explanation of the formation of meanders in a river has caused some problems for geomorphologists. In low flow conditions straight channels are seen to have alternating bars of sediment on their bed and the moving water is forced to weave around them. This creates alternating shallow sections (*riffles*) and deeper sections (*pools*). The swing of the flow that has been induced by the riffles directs the maximum velocity toward one of the bends, and results in erosion by undercutting on that side. An outer concave bank is therefore created. Deposition takes place on the inside of the bend, the convex bank. Consequently, although the river does not get any wider its sinuosity increases. Once created, meanders are perpetuated by the corkscrew movement of water called *helical flow*. This is a surface movement of water across to the concave outer bank with a compensatory subsurface movement back to the convex inner bank. Eroded material from the outer bank is transported away and deposited on the inner bank.

The cross-section of a meander is therefore asymmetrical. The outer bank forms a *river cliff* with a deep pool close to the bank, while the inner bank is a gently sloping deposit of sands and gravels, called a *point bar*.

As erosion continues on the outer bank, the whole feature begins to migrate slowly both laterally and downstream. In some parts of the world meanders regularly change their course, causing difficulties for those who determine land ownership boundaries.

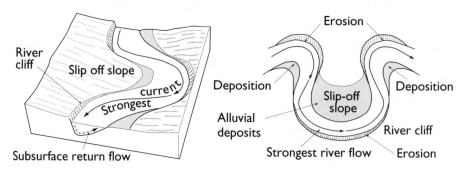

Meander

measures of central tendency are attempts to describe or summarize a set of data by calculating a single numerical figure. There are three main measures: *mean*, *median* and *mode*.

mechanical weathering: see *physical weathering*.

mechanization is the substitution of human or animal forms of power by a mechanical system. Examples include the use of tractors and combine harvesters in farming, the use of production lines and robots in factories, and the use of motor vehicles for transportation.

median: the middle value of a set of numbers arranged in order of size. In general terms, if there are "n" numbers in a data set, then the median is the $(n + 1)/2$ th value. It is one of the *measures of central tendency*. Unlike the mean, it is an actual value in the distribution of numbers, except where there are an even number of values, in which case the median is the average of the middle two.

Mediterranean climate: a feature of the Mediterranean area of Europe, California, central Chile, Cape Province in South Africa and parts of southern Australia. The main characteristics of the climate are:

- hot dry summers
- warm wet winters
- an annual temperature range of approximately 15°C
- an annual precipitation of 400–700 mm.

The summers are the result of the poleward movement of the *intertropical convergence zone (ITCZ)*, which causes the subtropical high pressure belt to migrate into these areas. Subsiding air and dry offshore winds bring arid conditions.

In the winter, the ITCZ and subtropical high move equatorward, and the area comes under the influence of westerly winds. These blow onshore and bring rain, often associated with depressions. Incursions of colder air from the poles also frequently occur, encouraging the further development of frontal rain systems.

California and central Chile have adjacent cold offshore ocean currents, and *advection fogs* are a regular event along the coastline.

The movement of different air masses into the Mediterranean region of Europe also causes local winds to arise. In the winter, cold air from the north is channeled violently into the Rhone Valley to create the mistral. In the summer, hot dry dusty winds such as the sirocco blow from the Sahara.

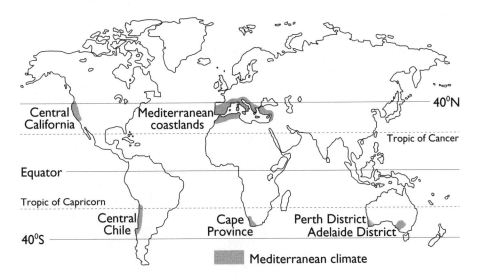

Areas with a Mediterranean climate

Mediterranean vegetation: the climatic *climax vegetation* associated with those areas of the world experiencing a *Mediterranean climate*. In all cases it is *xerophytic* or drought-resistant. The original vegetation of the Mediterranean areas of Europe was mixed forest of conifers and broad-leaved evergreens. However, because of human interference little of this remains. The characteristic vegetation is now a mixture of small trees, shrubs, low scrub and grassland. Local names for this type of vegetation include *maquis* and *garrigue* in the Mediterranean, and *chaparral* in California. Most plants in these areas have a range of adaptations to resist the summer drought:

- thick waxy evergreen (*sclerophyllous*) leaves to reduce water loss by *transpiration*, for example, the arbutus tree
- the ability to close their stomata, for example, the mastic tree
- very small leaves or thorns to reduce transpiration, for example, gorse
- deep root systems, for example, the almond
- bulbs and tubers that remain dormant in the soil and burst into life during the wet winters, for example, the hyacinth
- thick gnarled barks to reduce transpiration, for example, the cork oak and olive.

Many trees are also pyrophytes or resistant to fire. The occurrence of many such trees supports the view that the Mediterranean vegetation is essentially a *plagioclimax* – the outcome of the activities of humans.

meltwater: the water that issues from a *glacier* or *ice sheet*. Meltwater may be found on the ice, in the ice or below the ice, but most often beyond the ice margin. Meltwater gives rise to *fluvioglacial landforms*.

mental map: a map drawn from memory by an individual that indicates those aspects of an area that are familiar or of particular importance or interest to that individual. An example would be to show a journey from home to school, or to a leisure facility. A mental map is rarely drawn to scale.

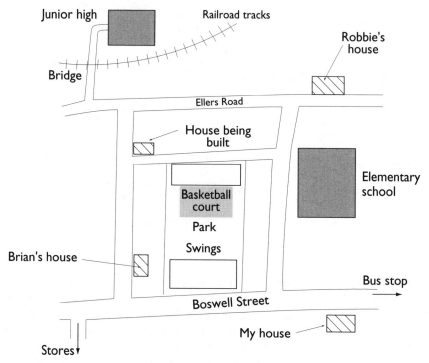

An 11-year-old's mental map

Mesolithic: the middle period of the Stone Age, which in Britain lasted from around 7500 to 3000 BC, extending from the Pre-Boreal through the Boreal and into the Atlantic climatic period. This culture, characterized by the use of small stone implements, probably moved into Britain from Europe before sea level rose to form the North Sea about 5000 BC. There is evidence of domestication of dogs and groups lived mainly by hunting, although artifacts from Star Carr, in North Yorkshire, suggest that this lakeside site was occupied by people who had boats and practiced fishing and hunting.

metamorphic rocks were either *igneous* or *sedimentary* in origin, but they have been subjected to alteration, becoming changed in both character and appearance. The original rocks may have been changed by increased temperatures or increased pressures or a combination of both. The intrusion of *magma* into a region may cause such a change, the rocks around the intrusion forming a metamorphic aureole. Metamorphic rocks are frequently crystalline, with the crystal aligned in one direction or divided into light and dark bands, and they rarely contain fossils. Examples of metamorphic rocks are slate, gneiss, schist and marble.

methane is natural gas, which is sometimes referred to as "dry gas." It is an oil-associated gas that can be produced from gas-rich oil fields or from separate gas fields. Methane is a mixture of hydrocarbons in a gaseous state at normal (ambient) temperature and pressure. This gas can be liquified by refrigeration to $-161.4°C$ when it takes up $\frac{1}{600}$ the volume of the gas. This liquified natural gas (LNG) is easier to transport – an important factor since there is considerable trade between the main producing areas and major markets, although the USA is both the world's largest producer and its largest consumer. Algeria, Saudi Arabia and the Persian Gulf states are major exporters; other significant producers are the United Kingdom, Norway, the Netherlands, Australia and New Zealand.

microclimatology: the study of climatic conditions and variations over a small area. These could include the changes in temperature, precipitation, humidity, wind speed and evaporation within a local climate such as a forest or on the shores of a lake. *Urban climates* contrast with those of their rural surroundings and different forms of vegetation or standing crop produce varied temperature and wind speed profiles at different heights above the ground.

mid-oceanic ridge: a range of submarine mountains formed at a divergent or *constructive plate boundary*. They occur mainly in the mid-Atlantic, the East Pacific, and the mid-Indian Ocean. They are formed largely of basaltic rocks and are associated with shallow foci *earthquakes*. The ridges can rise 3000 meters above the ocean floor and extend up to 4000 km wide; some locally rise above sea level to produce islands, e.g. Ascension Island. The form of the ridge crest is controlled by the rate of separation of the plates. New crust forms as the plates diverge when basaltic magma moves toward the surface. Where there is a slow rate of separation, 10–50 mm per year, as in the mid-Atlantic, there is a marked rift valley along the axis of the crest. It is 30–50 km wide and up to 3000 meters deep, giving inward-facing scarps with volcanic activity along the rift floor. Where there is more rapid spreading, 50–90 mm per year, as in the Galapagos ridge, the rift is less marked and the ridge appears smoother. With rapid separation, over 90 mm per year, as in the East Pacific Rise, the crest is smooth and unbroken. On either side of these ridges the ocean floor reveals alternate bands of reversed magnetism set into the rock as it formed. This form of *paleomagnetism* provides evidence of *sea floor spreading* and plate movement. The age of rocks increases with distance from the mid-oceanic ridge.

migration is a term used for any kind of movement; in the case of population movement it usually refers to a permanent change of residence. The United Nations defines "permanent" as a change of residence lasting more than one year. Recently the term has been applied more widely to include movements that are seasonal or even daily, but such temporary moves are perhaps more accurately described as circulatory movements. Migration can be classified as either voluntary or forced and a distinction can be made between internal migration, within a country, and external or international migration. Voluntary moves are more selective, i.e. only certain individuals or groups of people may be affected by a particular set of factors. Forced migration is less selective: generally the whole population is forced to move rather than being the result of individual choice.

Minamata disease: a disease resulting from mercury poisoning that arises out of industrial *pollution*. Paralysis and mental illness are the most common symptoms. The name comes from the town in Japan where the disease was first experienced. A petrochemical factory in the area had for years been dumping the waste from an acetaldehyde-making process into the local bay. As more people contracted the disease, research concluded that the waste contained mercury, which is highly poisonous, and this had been gradually passed into the population in small doses through the consumption of contaminated fish.

mineral deposit formation: the type of structure in which minerals are found. A mineral ore is an accumulation in sufficient concentration to warrant commercial exploitation and the formation or structure in which it is located determines the relative ease, and cost, of extraction. There are three main types of formation: igneous ore bodies, sedimentary ore bodies, and alluvial deposits.

As magma was pushed into the crustal rocks it cooled and different mineral constituents separated to form magmatic ore deposits such as magnetite (iron ore), copper and nickel, chromite and platinum. Liquids and gases were often forced upward through fissures to form veins and lodes in which different minerals solidified at different temperatures and accumulated at varying depths below the surface. These formations are generally difficult and expensive to mine; they follow irregular courses, their composition varies and they often end abruptly.

Sedimentary ores are found in more uniform, dipping or horizontally bedded layers, resulting from deposition on the floor of the sea or a lake. These include iron ores such as limonite, manganese and phosphates as well as evaporites such as gypsum, potash and rock salt. Some deposits result from the downward percolation of mineralizing solutions; bauxite and some nickel deposits are formed in this way. These formations are much easier to exploit, particularly if the "overburden" (the geological deposit covering the sedimentary ore) is poorly consolidated material such as glacial drift. These deposits can be mined by opencast methods.

Where veins, lodes and sedimentary formations have been subjected to erosion, minerals may occur as deposits in sands, silts or clays along valley floors or on the continental margins in shallow oceans. These are termed placer deposits and include tin, gold and diamonds. They can be worked by pumping or dredging, a relatively inexpensive form of mining, because the minerals have already been removed from their natural setting by erosion.

mining is the extraction and preparation of minerals for industrial use. The commercial exploitation of a reserve is influenced by a number of factors. Geological considerations determine the ease of accessibility to the mineral body, reflecting the *mineral deposit formation*. Other factors such as the depth of the deposit influence the choice of extraction method. Shafts and galleries associated with lodes and veins are more expensive than opencast and dredging techniques. The quality, or metal content, of the ore reserve is also relevant to the decision to exploit a mineral. High content can offset the high costs of exploitation, as in Kiruna and Gallivare (north

Sweden) where the ore is 55–70% iron. The Lorraine ores are of 24–35% content; these need to be mined near to their market or where they are easily removed from the sedimentary structures. The size of the reserve is also important as this influences the potential lifespan of the activity; a large reserve can adequately return the investment required to develop the mine.

Economic factors such as the availability of capital for investment, a local labor force (or funds to attract an outside workforce) and the accessibility of the mining area in relation to developed transportation infrastructure are also influential. Political considerations such as the strategic value of reliable mineral supplies and the political stability of the mining region may also influence potential investors.

mist is a suspension of water droplets in the air, formed in a similar way to *fog*. It is an atmospheric condition where visibility is reduced but is greater than 1 km.

mixed economy: a combination of a *market economy* and *centrally planned* enterprises. *Market-led* economies have certain perceived disadvantages, the most important of which is the need for governments to regulate the workings of the market either through laws or by running parts of the economy through state enterprises. This is known as a mixed economy.

Centrally planned economy	Mixed economy	Market economy
State runs and organizes production	Some production run by state, e.g. utilities	Market-led production
Labor works for state	State and private sector compete for labor	Wages determined by market forces
Prices controlled		Prices determined by supply and demand

mode: the value that occurs most frequently in a set of data. It is a *measure of central tendency*.

models are simplified representations of the real world that are used to describe and analyze general situations. By making general assumptions about the real world researchers can focus their attention on one particular factor. For example, Johann Heinrich *von Thunen* (land use zonation around an urban market), Alfred *Weber* (industrial location) and Walter *Christaller* (central place theory) all assumed that they were dealing with a "uniform plain," i.e. a landscape in which background factors such as climate and soil fertility, and production factors such as land and wage costs, are uniform. They also assumed that man behaves as "economic man," a rational decision-maker who has perfect knowledge of all costs and market demands and who aims to maximize profits as an "optimizer." The purpose of making such unrealistic assumptions was to allow each researcher to study the effect of transportation costs on land use, industrial location or service provision without having to initially deal with the whole spectrum of factors. Once these had been investigated in a model landscape they could be compared with the real world and other modifications could be introduced. In this way the model acted as a "control" against which real situations could be assessed.

Mathematical models can also be developed to describe the relationship between physical and/or human factors, for example in analyzing stream discharge in relation to inputs of precipitation. Simulations using chance elements can be used to predict changes, such as the stochastic (chance) models applied in *Monte Carlo* techniques.

Physical models may be constructed as small-scale working models to represent reality; the influence of sediment load and channel characteristics can be investigated under laboratory conditions.

Mohorovicic, Andrija: discovered the junction between the Earth's *crust* and the *mantle* in 1909 by studying the different speeds of *earthquake* waves as they passed from the crust to the mantle. This boundary is termed the Mohorovicic discontinuity, or Moho, and it lies at a depth of up to 40 km beneath the continental areas and between 6 and 10 km below the ocean floors.

monoculture occurs where agricultural production relies on one dominant crop. This may take the form of *extensive agriculture*, as with wheat growing in parts of the American prairies, or may occur with *intensive* systems, such as rice cultivation in Southeast Asia. A concentration on one crop can cause problems with mineral depletion in the soil leading to infertility, and for commercial producers the economic success of this type of system is very dependent on annual yields and the effect on market price. Overproduction in high-yielding years can lead to a low return as a result of low prices on the world market.

monopoly exists, in theory, when a single producer controls the supply of a particular product to a market. In practice such complete control rarely exists, but if one organization controls a sufficiently large proportion of the market it can exert control over prices by restricting output. The danger with monopolies is that they can exploit the consumer, charging higher prices or offering a poor service, or they can waste resources through inefficiency due to lack of competition.

monsoon refers to a seasonal reversal of wind direction. The best-known monsoon, in India and Southeast Asia, occurs because:

- in June the *intertropical convergence zone (ITCZ)* moves north but extends itself further north over northern India because of the intense heating that takes place there. This low pressure draws in warm, moist, unstable air from the Indian Ocean, bringing heavy rain. Rainfall totals are further increased by the uplift of the air over the foothills of the Himalayan mountain range, and by intense convection
- in January the ITCZ moves south over the equator. At the same time the center of Asia experiences intense cooling and a large area of high pressure develops. Winds blow away from this high pressure cell, bringing very dry conditions to the majority of the Indian subcontinent. The eastern coast of India and Sri Lanka may also get rain from this monsoon as the winds pass over the Bay of Bengal and may become moist.

The occurrence of the wet monsoon in June is unreliable because upper atmospheric conditions have to be right to allow the winds to come from the south. Failure may result in prolonged drought. Conversely, excessive rains may cause serious flooding.

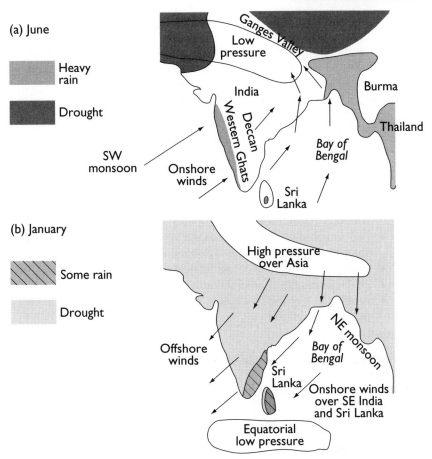

(a) June

Heavy rain

Drought

Ganges Valley
Low pressure
India
Burma
Deccan
Western Ghats
SW monsoon
Onshore winds
Bay of Bengal
Thailand
Sri Lanka

(b) January

Some rain

Drought

High pressure over Asia
Offshore winds
Bay of Bengal
Sri Lanka
NE monsoon
Onshore winds over SE India and Sri Lanka
Equatorial low pressure

The Indian monsoon

Monte Carlo simulation uses random numbers to determine, or simulate, the development of a geographical pattern across a landscape. For example, the spread, or diffusion, of settlement away from the initial point of growth could be simulated using this method. If a grid were established around the initial point of growth, each square in the grid could be given a weighting according to its chances of being settled. This weighting could reflect background factors such as accessibility, height, slope angle, water supply, resource availability and so on. This grid, the probability matrix, determines the chances of a particular square being selected for settlement; numbers are allocated to each square to reflect its weighting. Thus a square that has a lot of positive attractions might carry a weighting of 25 and be assigned 25 numbers, while a less attractive square could be assigned 2 numbers. When random numbers are selected, the squares in the matrix with a larger allocation of numbers have a greater chance of being chosen as points for settlement. However, because of the stochastic, or chance element, the random selection could lead to low-scoring squares being settled; this allows for less predictable chance factors to influence diffusion. The spread is influenced by chance elements set within certain constraints that reflect the operation of geographical factors. This type of simulation has been applied to the diffusion of ideas, as in the spread of hybrid seeds or farming equipment through an agricultural region, as well as the diffusion of people.

mor is an acidic form of *humus*. It is common in wet and cold environments and is associated with *heathland* areas and *coniferous woodland*. Mor is dark brown to black in color, poorly decomposed and lacking in nutrients. Few species of soil fauna can tolerate its acidic conditions and thus earthworms are rare.

moraine is the collective name for the debris that a *glacier* transports and eventually deposits. The debris has either been eroded by the glacier along its sides or base, or has fallen onto it from rock faces above. There are several types of moraine:

- lateral moraine – material found along the edges of the glacier
- medial moraine – formed by two lateral moraines merging when a tributary glacier meets a main glacier
- end moraine – material deposited across the front of the glacier. There are a number of variations of end moraines – see *terminal moraine*
- supraglacial moraine – material resting on top of the glacier
- englacial moraine – material within the glacier itself
- subglacial (ground) moraine – material carried along the valley floor beneath the glacier.

(See also *lodgment* and *till*.)

morphology refers predominantly to the shape of a feature. The morphology of a physical landform is its relative size, shape and field location. The morphology of a settlement refers to its shape in plan as well as to the degree to which its buildings are grouped together.

mountain and valley winds: see *anabatic* and *katabatic*.

mountain building: see *orogenesis*.

mountain climates are generally cooler and wetter than lowland areas of the same latitude. Temperatures fall by 1°C per 150 m rise, although the decrease in pressure, humidity and the amount of dust particles means that more shortwave radiation penetrates to ground level. This gives high daytime surface temperatures on slopes facing the sun. Poleward facing slopes, and those shaded from the sun, receive much lower *insolation*. Differential heating of air on valley sides and valley floors during the day may lead to warm air rising upslope, or up valley, as *anabatic winds*. As higher areas cool down, the cold dense air may descend into the valley floor, or move down valley as a *katabatic wind*. As moist air passes over mountains it is cooled to *dew point*, producing condensation and relief, or orographic, rainfall. This tends to increase with altitude up to about 2000 m above sea level, but above this altitude the low moisture content of air due to low temperatures produces a decrease of precipitation with height. As air descends after crossing a mountain barrier it is warmed adiabatically, producing the *Fohn effect*.

mudflow: a form of *mass movement* in which debris of varying sizes may be transported in a matrix of saturated clay. These are more rapid flows, which can occur on relatively low slope angles compared with *earthflows*. They occur in areas that experience torrential rain falling on ground that has limited protection from vegetation cover. This allows the weathered regolith to become saturated, increases the pore water pressure in the debris and reduces the frictional resistance between particles. As a result the debris moves downslope.

mull is a mild *humus* that is soft, black in color and rich in nutrients. It is produced by the action of bacteria and earthworms when the soil is not too acidic and is commonly found beneath lowland deciduous hardwoods and grasslands.

multicultural society: a social grouping that contains members from a wide variety of national, linguistic, religious or cultural backgrounds. It is often an emotive issue, especially when "cultural" differences are interpreted as racial differences. Although many modern groups are thought to have descended from three broad racial types – negroid, caucasoid and mongoloid – the distinctions are now so blurred that race has little scientific relevance. Although skin color remains as a visible distinguishing feature, people differ from one another in terms of ethnic differences, language, religion and culture. Multicultural societies are the result of *migration*, both forced and voluntary moves, but may also generate movement as persecuted minorities seek to escape oppression. Groups within a multicultural society may be integrated to a greater or lesser degree; the operation of *apartheid* in South Africa illustrated a lack of integration while Singapore represents a more tolerant attitude between the dominant Chinese (76%), Malays and Indians (22%), and the minority Europeans and Eurasians.

multilateralism: the development of trading agreements and negotiations between groups of countries, as distinct from bilateral talks, which take place between two parties. Negotiations may be undertaken on behalf of a trading bloc such as the *European Union*. An example of this type of agreement was GATT, the *General Agreement on Tariffs and Trade*.

multinational: see *transnational*.

multiple hazards: areas that are prone to *hazards* in general. The Los Angeles area comes under this heading, as it suffers from a range of hazards:

- physical environment – *earthquakes*, coastal storms, river and coastal *flooding, landslides, drought, smog* and bush fires
- human environment – racial riots, high crime levels, gang warfare and tanker spillages.

multiple-nuclei model: see *Harris-Ullman model*.

multiplier effect: a new or expanding economic activity in an area creates extra employment and raises the total purchasing power of the population, which in turn attracts further economic development creating more employment, services and wealth. This has been described as a case of "success breeds success." The multiplier effect is part of the wider process of *cumulative causation*, put forward as a model by Gunnar *Myrdal*.

multiracial society: see *multicultural society*.

Myrdal, Gunnar: a Swedish economist who in the 1950s first suggested the idea of *cumulative causation* and the *multiplier effect*.

NAFTA: see *North American Free Trade Area.*

nationalism: loyalty and devotion to the state, such that national interests are placed above individual or global interests, i.e. the assumption that nations are the primary focus of political allegiance. It can be present in states of widely diverse types.

nationalization describes the transfer of firms and industries from the *private* to the *public sector.*

national parks: areas of outstanding scenery where human activity is carefully controlled and where the environment is preserved or improved for the enjoyment of people and the *conservation* of native plants and animals. National Parks exist in many parts of the world, and there are slightly different criteria for their establishment and operation from country to country. The first National Park was Yellowstone, established in 1872.

natural increase: the increase in population within a country calculated by finding the difference between *birth* and *death rates.* If in one year, for example, the birth rate was 30 per thousand and the death rate 12 per thousand, then the natural increase rate would be 18 per thousand or 1.8%.

natural resources are those features that are needed and used by people and include such things as climate, soils and *raw materials* (iron ore, coal, crude oil, etc.). Natural resources form a category distinct from human resources, the latter of which includes such features as people and *capital.*

nature reserve: an area set aside by the government for the purpose of preserving certain plants, animals, or both. It is distinct from a *national park*, in that the park protects land and wildlife partly for public enjoyment, whereas a nature reserve protects animals (particularly birds) and plants for their own sake. Endangered species, for example, are often kept in nature reserves that protect them from hunters. The idea of protecting animals to keep them from dying out arose in the 19th century.

neap tide: a tide with a low range; the high tide is relatively "low" and the low tide is relatively "high." These occur twice a month, coinciding with the first and last quarter of the moon. At this point the sun, moon and earth are at right angles, with the Earth at the apex. This reduces the tide-producing pull of the sun and moon on the waters of the Earth because the gravitational forces between the three bodies are in opposition. This position is referred to as "in quadrature." With *spring tides* the bodies are in line, producing a much stronger combined gravitational effect to create a high tidal range.

nearest neighbor analysis is a precise mathematical means of measuring point patterns. The technique involves using a simple formula, giving an index that shows how far the measured pattern of settlement, for example, is from being "random," i.e. generated by chance. Greater or smaller indices indicate that the pattern has a tendency to be either spaced or clustered. The index value runs from 2.15 (evenly spaced) through 1.0 (random) to zero (highly clustered).

FORMULA $\qquad Rn = 2\bar{d}\sqrt{\dfrac{n}{A}}$

where $\bar{d}$ = the mean distance between nearest neighbors
$\quad$ n = number of points
$\quad$ A = area in question

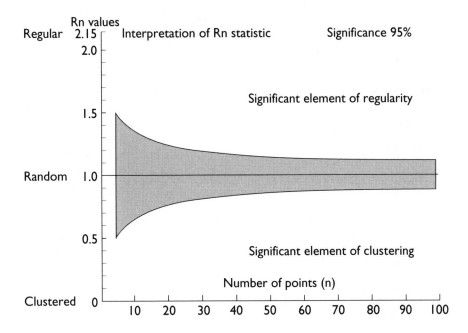

negative correlation: a *correlation* where an increase in the value of one *variable* is matched by a decrease in the value of the other variable. For example, a correlation between *birth rate* and the *GNP* of countries would be negative, as generally the higher the birth rate of a country, the lower the GNP.

negative feedback is when a *system* acts by lessening the effect of the original change and ultimately reversing it.

On a river system, for example, increased erosion may cause so much debris to accumulate within the valley that the river cannot move it, which will result in deposition. As a consequence, there will be less downcutting and the prevention of slope steepening. Thus, the original increase in the erosive capacity of the river has turned into a period of deposition.

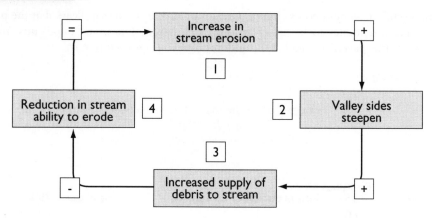

Negative feedback

negative skew is a bias within a *distribution* toward high values.

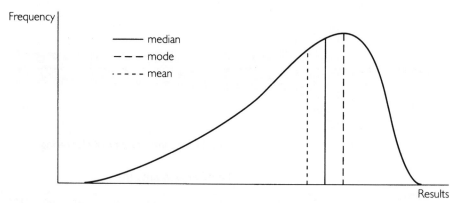

Negative skew

neighborhood watch: schemes in which the public and police combine in an attempt at crime prevention. The aim is for alert neighbors to deter crimes of opportunity, such as the theft of an unlocked car, *burglary* of an unprotected house, and crimes against the person. In return, the police give advice on domestic security methods. The most enthusiastic neighborhood watch groups have been in the *inner cities*, where the main victims of crime live.

neocolonialism: the acquisition or retention of influence over other countries, especially one's former colonies, often by economic or political measures.

Neolithic period: a cultural period that followed the *Mesolithic* and extended until the *Bronze Age*, i.e. from about 3500 BC until 2000 BC. These groups used flints and polished stone tools, and artifacts indicate that they cultivated crops, domesticated animals and made pottery. Evidence comes from earthen and stone tombs found widely throughout Britain; tapering burial mounds were constructed as well as sites such as Stonehenge and Avebury, which indicate a higher degree of organization.

net primary productivity (NPP) is the amount of organic matter that is available for man and other animals to harvest or consume. Gross primary productivity is the rate (in dry grams per square meter) at which matter builds up in green plants as a result of photosynthesis. As plants respire some of this is lost and NPP is the gross PP minus the amount lost due to plant respiration. World variations in NPP are due mainly to differences in moisture and temperature, and the availability of nutrients. There are considerable variations from 2200 dry $gm/m^2/year$ in tropical rainforest, 1250 in temperate forests, 900 in savanna areas to 90 dry $gm/m^2/year$ in desert and semidesert areas. The average marine value is 152. The stage of plant succession also influences the level of NPP. On land nutrients are stored in the soil; in sea areas this lack of nutrient supply is the greatest limitation on NPP. Nutrients are lost due to sinking in the deepest oceans to depths below the limit of light penetration. The most productive parts of the oceans are on the margins and the continental shelf, or where upwelling water brings nutrients to the surface.

network: a number of places joined together by links to form a *system*. The links are known collectively as the *infrastructure* and could be such features as roads, railroads, canals, sewage and telephone systems. To allow comparisons to be made between networks it is necessary to adopt a universally acceptable method of description. For many systems in network analysis it is usual to represent the network as a series of straight lines. Among some of the common methods of analysis are the Shimbel Index, the alpha, beta and gamma indices, the cyclomatic number and the detour index.

newly industrialized country (NIC): countries that since the 1960s have shown rapid growth in *manufacturing industry*. The most rapid growth has occurred recently in the *Asian "Tigers"* of Singapore, South Korea, Taiwan and Hong Kong. Two main aims are common for such rapid industrialization:

- import substitution – manufacture of goods previously imported for sale on the home market
- export promotion – sales to the world market helped by the creation of *free trade zones* and export processing zones.

new technology is used to describe the rapid changes in communication and other processes resulting from the exploitation of the silicon chip. The *innovations* resulting from the use of new technology can be split into three main types:

- by process, that is, advancements in *manufacturing* technology and automation
- by product, that is, new product opportunities using microelectronic technology such as the fax machine and electronic games
- by communication links, that is, *information technology*.

NIC: see *newly industrialized country*.

niche: the role a single organism or plant plays in an *ecosystem*. It may refer to the organism's place in a *food web*, or to a precise description of a plant's *habitat*.

A niche glacier is a small upland body of ice resting on a slope or in a shallow hollow. They are common on north-facing slopes.

NIMBY ("not in my back yard") is applied to situations where a proposal may be wanted by a group of people for a common good, but no one is prepared to have it near to where he or she lives. Examples of such situations include the building of new highways or the development of a potentially hazardous industry. Most people agree that these are either useful or necessary. However, strong resistance from those affected most by their construction and use is a regular feature of such schemes. Often the area with the least powerful "voice" is chosen, and sometimes more than once, leading to a concentration of negative developments in an area.

nitrates are nutrients that are essential for plant growth. The main sources of nitrates are the soil itself, fertilizers, animal manures and silage liquor. A popular misconception is that the high levels of nitrates in groundwater and in some rivers and lakes are due to excessive applications of nitrogeneous fertilizers that find their way into the water. Research has shown that relatively small amounts of nitrates are left over from fertilizers to be leached away. However, levels of nitrates in groundwater, rivers and lakes are increasing, and in extreme cases cause *eutrophication*. Nitrates are produced naturally by microbe activity in the autumn as they break down organic material after a harvest. Methods that have been suggested to farmers to reduce the amount of nitrates being leached out by the autumn rains include:

- not using fertilizer during this time
- not leaving land bare as plowing causes a surge of microbe activity and therefore nitrogen release
- sowing winter crops early so that roots can take up the nitrates and thereby reduce leakage.

nitrogen cycle: *nitrates* that exist in soil are taken up by the roots of plants, for example grass. Herbivores, such as cattle and sheep, eat the grass and then release the nitrogen as ammonia by excretion. The excreta decays on or in the soil and the ammonia is converted by bacteria into nitrates, which are then leached back into the soil. There are additional inputs of nitrogen from the atmosphere, as well as outputs via seepage into *groundwater*.

nivation: the process by which a hollow containing a snow patch becomes wider and deeper. *Freeze-thaw action* beneath the snow patch causes the underlying rocks to disintegrate. During warmer periods meltwater from beneath the patch flows away from the hollow and washes away the weathered material. In time the hollow may enlarge so that it contains sufficient snow to last through a summer. When this is the case, the hollow becomes an embryo *cirque*.

node: a point in a transportation network representing a town, station or junction within it. On a *topological map* a node is represented by a dot, and is alternatively known as a vertex.

nomadic: a description of an existence that involves a group of people roaming from one place to another seeking pasture or water for their animals. Nomads do not have a permanent home. Nomadic herding usually exists in areas where the climate is too extreme to support permanent settled agriculture. The climate may be too arid, cold or hot, and the vegetation is sparse. Nomads will tend to follow seasonal rainfall, or will move away from snow-covered grazing grounds. Examples include the Tuareg of the Sahara, the Rendille of northern Kenya, and the Lapps of northern Finland.

In some parts of Africa, nomadism is being actively discouraged. *Overpopulation* and *overgrazing* have caused resources to be overstretched. Nomads are being forced to move near to small towns and to become sedentary farmers. Wells and other services have been provided. However, there is a reluctance by the nomads to part with their animals, and this has led to extreme forms of overgrazing around the settlements.

nongovernmental organization (NGO): a body that has not been created by law but appears to have major responsibilities. They are created mostly by volunteers, but are later supported by individual or corporate benefactors. Many are charities, and have international responsibilities such as organizing relief and humanitarian operations, as in the case of the Red Cross. Others are national in their outlook and may have environmental concerns. *Greenpeace* is an example of a NGO with significant worldwide following and influence. Frequently NGOs are given the responsibility of administering funds from governments, as they are seen to be impartial.

nonrenewable resources are those that are finite, as their exploitation can lead to the exhaustion of supplies. In the energy field, coal, oil and natural gas are classified as nonrenewables.

nontariff barriers are hidden barriers to trade imposed by governments because they wish to restrict imports without being seen to do so, perhaps because it would be contrary to international regulations under *GATT*. Such barriers may take several forms:

- constantly changing technical regulations, which makes compliance difficult for importers
- forcing importers to use specified points of entry where documentation is dealt with only slowly
- regulations that favor domestic producers, e.g. packaging, and labels that conform to local language requirements.

normal curve: the bell-shaped curve of the *normal distribution* as shown below.

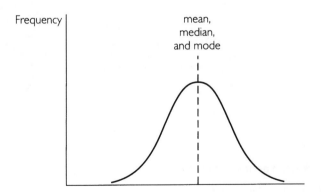

Normal distribution

normal distribution: a *frequency distribution* that is bell-shaped so that half of the variables lie to the left of the *mean* and half to the right. Due to the symmetry of the curve, the mean, *median* and *mode* all coincide at the same value. If data is collected on an event that occurs many times over, variations in the result will tend to form a pattern that is known as normal distribution. This means that the results will tend to be clustered around the average, and will be split equally between results above and below the average. Normal distributions can be standardized so that the width of the distribution is about six *standard deviations*. The areas within each standard deviation can be very useful in statistical analysis, for example in *significance testing*.

North: a term used to indicate the wealthier nations of North America, Europe, the former USSR, Japan, Australia and New Zealand. This category therefore contains the countries of the *First World* and the former *centrally planned economies* of the *Second World*. The term *economically more developed countries (EMDC)* is now often used for this group. The term "North" first appeared in the *Brandt Report* (1980). (See also *South*.)

North American Free Trade Area (NAFTA), whose members consist of the USA, Canada and Mexico, is an association formed in 1992 to create an area of free trade in North America. Over time tariffs and other trade restrictions between member countries are being abolished. After the *single European market* NAFTA is now the largest free-trade area in the world.

North/South divide is the line that separates the richer countries of the *North* from the *economically less developed countries (ELDC)* of the *South*. The difference in economic development between the two areas is known as the *development gap*.

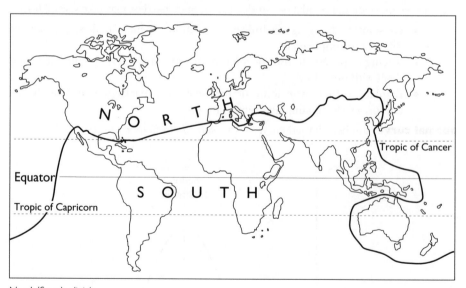

North/South divide

Not In My Back Yard: see *NIMBY*.

NPP: see *net primary productivity*.

nuclear energy involves the use of radioactive energy produced by nuclear reaction. The heat released by the reaction is used to produce steam, which drives turbines to produce electricity. Uranium is processed, enriched and converted to uranium dioxide, which is used in the reactor. This undergoes *nuclear fission*, releasing large amounts of heat. Only small amounts of uranium are needed to produce a given output of heat compared with other forms of fuel. The plants are expensive to construct and decommission, and there is a major concern relating to the disposal of the *waste* material, which is a problem because it is radioactive and requires specialized treatment. The public is worried that some contamination will occur in transit or at the point of disposal. However, this must be seen in the context of world energy needs: it is unlikely that demand can be met from renewable sources and yet other sources such as coal, oil and gas are finite. There is no shortage of uranium and operating costs are very competitive. However, there are questions of safety and environmental damage, but so far nuclear plants have a good record in relation to other forms of electricity production and they produce less CO_2 and SO_2. Nevertheless, an existing public fear is exploited by nuclear opponents such as *Greenpeace*.

nuclear fission is a reaction in which a heavy nucleus splits into two parts that then emit neutrons, releasing energy in the process. The fission of uranium releases two or three neutrons that are able to split more atoms; thus a chain reaction is established that releases energy.

nuclear fusion involves joining light atomic nuclei with heavier nuclei. This fusing together of different atomic masses creates energy without any loss of mass. This was achieved for the first time in 1991 at Culham Laboratories, a Research Establishment near Oxford in England.

nuclear waste includes both high-level and low-level radioactive material. Power stations produce high-level waste consisting of used fuel rods and cells that have been removed from the reactor. These are transferred to a reprocessing plant, where reusable uranium and plutonium are separated out to leave radioactive waste. Low-level waste includes clothing and materials used in hospitals where exposure to radium and X rays can present a low-level risk. These materials are disposed of in controlled chemical and radiation dumps and may be buried in suitable geological structures.

nucleated: a term applied to a settlement in which the buildings are grouped around some focal point such as a church, a park or a water supply. Historically this may reflect the search for a wet point site in a porous rock area, or the need for a dry point site in a wet landscape. It could be a defensive point or a central location in an area of *open fields*. Nucleated settlements display various forms: linear, T-shaped or cruciform, or they may reflect the shape of the park: circular, rectangular or lens-shaped.

nuée ardente is a glowing cloud of hot gas, steam and dust, volcanic ash and larger *pyroclasts* produced during a violent eruption. This descends the slopes of a volcano at high velocity, as in the case of the eruption of Mt. Pelée on the Caribbean island of Martinique in 1902, giving rise to the term *pelean eruption*.

null hypothesis: a negative assertion that states there is no relationship between two variables that are being tested, e.g. "there is no relationship between building height and distance from the center of the central business district." It assumes that there is a high probability that observed differences between two sets of data are due to chance variations. If the null hypothesis can be rejected statistically, then we can assume that any differences between the data sets are not due to chance but are the result of differences between the two statistical populations. We can then suggest that the *hypothesis* is acceptable.

nunatak: a rocky mountain peak that projects above an ice sheet. Because it is not covered by ice the peak is not planed off or rounded and *freeze-thaw* action maintains the steep sides. After deglaciation, the jagged and angular peak contrasts with rounded surfaces that were covered by ice.

nutrient cycle: the circulation of minerals around the *ecosystem*. Nutrients are taken up by the root system of plants. These are then used and released as the plants shed organic matter such as leaves, or when the plant dies. Some nutrients pass along the *food chain* as herbivores consume leaf material. These return minerals to the system through excreta or on their death. The litter that accumulates on the surface is broken down by microorganisms and fungi that return nutrients to the soil store to be used in the cycle again. The three main storage areas within the system are the litter layer, the soil and the *biomass*. The relative importance of each of these stores depends upon the nature of the *biome*. In *tropical rainforest* areas, a very high percentage of the nutrients are stored in the biomass, which consists of several layers of plants. High temperatures and rainfall produce rapid decay of the litter layer and nutrients are quickly returned to the soil from where the many plant roots take up minerals. In *coniferous forests* the litter is the dominant store. Lower temperatures restrict breakdown and there is a slow return of nutrients to the soil. This slow replacement is insufficient to make up for the loss through *leaching*. The biomass is a relatively low proportion of total nutrients; it is made up of one main tree layer, consisting of species that have thin needle-like leaves, and the acidic litter restricts the lower layers of vegetation.

nutritional requirement: the amount of food needed to sustain an individual. The daily intake of calories necessary to sustain a person averages about 2200 calories per day but this must be seen as a very general figure because an individual's need will reflect a variety of factors: climate, level of activity, type of employment, body size. It is not simply a measure of calorific intake; proteins, energy foods, vitamins and minerals are essential elements of a diet. If these needs are not met then *malnutrition* may result. Carbohydrates, which provide energy, are critical, and the intake of these is generally the measure used to show differences in diet, expressed as calories/person/day.

oasis: an area in the middle of an *arid* region that is made fertile by the presence of water. Oases are usually small, but they can cover over a hundred square kilometers. The reason for the presence of water is usually the occurrence of water-bearing strata (an *aquifer*) on the surface at that point.

obsolescence occurs when a product, service or machine has been overtaken by a new idea that provides the same function in a better or more attractive way. This often occurs as a result of new technology, though it may also be due to changing lifestyles, fashions or scarcity of resources. The threat of obsolescence encourages most firms to update their products regularly and to look for new products that can replace declining ones. Some firms look to gain financial advantage from two specific types of obsolescence:

- built-in obsolescence means that the product has been designed to last only a limited amount of time
- planned obsolescence is the creation of a feeling on the part of customers that they should replace items that are in fact usable. The car industry is a good example, where the addition of new features to existing models by manufacturers means that the customer is persuaded of the need for change.

Both these types of obsolescence are regarded by environmentalists as undesirable ways of using the Earth's resources.

occlusion: in a mid-latitude *depression,* when the cold front catches up with the warm *front,* this then forms an occluded front. If the air behind the warm sector is colder than the air in front of it, a cold occlusion occurs, and if the air is warmer, a warm occlusion.

Cold occlusion

ocean basin: an area of ocean beyond the edge of the *continental shelf.*

ocean currents: large-scale movements of water within the oceans that are part of the process of the horizontal transfer of heat polewards. Other currents carry colder water toward the tropics. In *temperate* regions, warm ocean currents generally accentuate the moderating effect of the oceans, particularly where the *prevailing winds* are *onshore.* The influence of the North Atlantic Drift on Western Europe is a good example. Ocean currents are largely set in motion by the prevailing surface winds associated with the *general atmospheric circulation.*

ocean thermal energy conversion is a method of exploiting the temperature difference in *tropical* waters between the warm surface of the ocean and the colder water at depth. This could be a form of *renewable energy*, which it is claimed would be one of the world's most environmentally benign. A number of small experimental schemes have been established off India and Taiwan, for example.

ocean trench: a narrow, deep depression in the ocean floor that corresponds with the *subduction* zone associated with convergent margins or *destructive plate boundaries.* The Marianas Trench, on the western margin of the Pacific Ocean, extends to depths below 11,000 meters. They are typically arc-shaped and correlate with the location of deep-focus *earthquakes,* although intermediate and shallow earthquakes are also formed on the margin of the trench toward the line of volcanic islands that form the *island arc.* Examples include the Chile-Peru Trench where the Nazca Plate (oceanic) subducts beneath the South American Plate (continental) and the Japan Trench where the Pacific Plate (oceanic) subducts beneath the Eurasian Plate (continental).

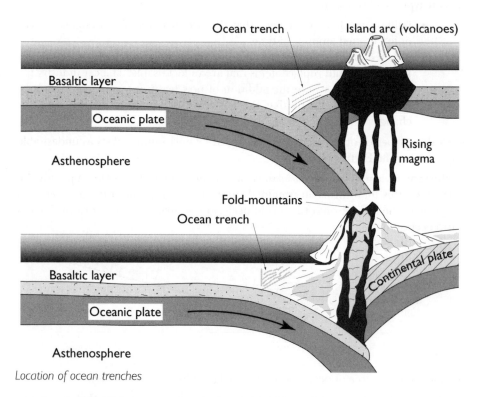

Location of ocean trenches

OECD: see *Organization for Economic Cooperation and Development.*

offshore: this term may be used in a number of ways:

- referring to movement from the land toward the sea; an offshore wind can develop at night because the sea retains its heat while the land radiates heat rapidly, causing cooling. The relatively warm air over the sea rises and is replaced by cooler air moving off the land to create a land-sea breeze

- in relation to coastal geomorphology, offshore features develop on the seaward side of the wave breakpoint, e.g. offshore bar
- in manufacturing many *transnationals* have located "offshore," i.e. they assemble the final product in a less economically advanced country using components produced in advanced industrial nations. Many American companies have taken advantage of cheaper labor and land costs in northern Mexico; the products are generally hi-tech, high value/weight ratio goods that can withstand the transportation cost incurred in moving components into the assembly area and in moving products back to an advanced market. Location "offshore" can also help a company to penetrate the local market.

oligopoly exists when a small number of companies dominate the market for a particular product, thus accounting for a very high percentage of output, e.g. car production. These firms tend to imitate each other's behavior in terms of model range and development in order to preserve their share of the market. This can appear to be a competitive situation, but this competition is usually in the form of special offers rather than price reductions as there is often a degree of collusion that maintains price levels and profits.

onshore: a sea to land movement; during the day heating of the land causes warm air over the land surface to rise. This is replaced by cool air moving in from over the sea creating a sea-land breeze.

OPEC: see *Organization of Petroleum Exporting Countries.*

opencast mining: a method of extraction that involves extensive excavation to remove the overlying material (overburden) covering the mineral deposit. This method is used particularly for working coal or iron ore where the mineral has been laid down in relatively gently dipping or horizontally bedded geological structures as in a sedimentary *mineral formation.* Large-scale earth movers and drag cranes are used to expose and extract the mineral. It is a lower cost form of mining, which can allow even relatively low-quality deposits to be commercially exploited, particularly where the overburden is easily removed, as in the case of unconsolidated glacial material in the iron ore "ranges" of the Lake Superior field.

open field system was a farming system introduced to the British Isles by Anglo-Saxon settlers and it continued to exist in some parts of the country until well into the 18th century. It consisted in its simplest form of two or three large fields around a central village, each lying fallow in turn. As more woodland around the village was cleared the fields became progressively larger with time. Each field was divided into a series of strips, usually one-third or one-half of an acre in size, and each family in the village held a number of strips scattered about in the fields. The farming of the strips was uneconomic in that land was wasted by their boundaries, and time was wasted in traveling to and from them.

The common width of the strip was 22 yards, and in many cases the length was 220 yards, or one furlong. The continuous plowing of these strips threw the soil into the center to create a slight ridge, separated by a depression or furrow. Each strip was separated by a double furrow, or a grass ridge, or balk.

During the late Middle Ages, the open field system began to be replaced by *enclosures*, and by the end of the 18th century it had disappeared in all but a few locations.

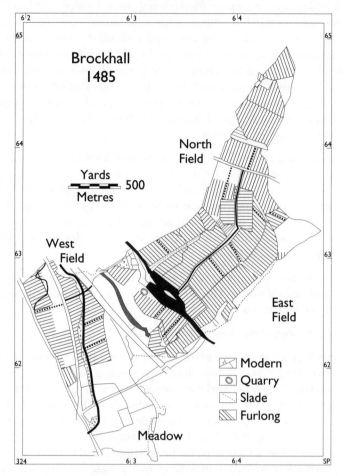

An example of an open field landscape

optimizer: a *decision-maker* who always aims to maximize returns, or profits, in any situation. It is an assumption that underlies a number of *models* based on the concept of *economic man*. An optimizer is assumed to have perfect knowledge of all the relevant factors that influence costs and revenue in arriving at a rational decision. It contrasts with the behavior of *satisficers*.

optimum population is the theoretical number of people who, when working with all the available resources, will produce the highest standard of living and *quality of life*. This is a dynamic situation, because as technology improves, new resources may become available, which will mean that a higher optimum population can be carried. If there are too many people for that living standard to be achieved, then the area is said to be *overpopulated*. Some areas have resources that are not fully exploited with the present population and therefore could stand a population increase. Such areas are said to be *underpopulated*.

order refers to goods and services, in that low-order items are obtained frequently and cost little per unit (bread, newspapers), whereas high-order items are more expensive and bought less frequently (furniture, clothing). People are prepared to travel much greater distances, therefore, to obtain higher-order goods and services. This means that low-order functions (convenience store, elementary school) are spaced much closer together than high-order functions (high school, hospital).

ordinal data is presented in terms of the relative importance (or order of magnitude or rank) of data rather than absolute values. Data can be classified into broad groups, e.g. sediment size given as fine, medium or coarse, or settlements placed into general groups based upon size.

organic farming is a form of agriculture that does not use chemical *fertilizers* or *pesticides/insecticides/herbicides*, but instead favors animal and green manures and mineral fertilizers such as fish and bone meal. The farming system will be *intensive*, but will involve crop rotation and the use of fallowing periods.

Pros:
- the increased *organic matter* in the soil enables it to retain more moisture during dry periods and allows better drainage during wet spells
- less likely to cause *soil erosion* or exhaustion, as apart from more organic matter, it will have more soil *fauna* (worms, etc.)
- no harm to the environment as there will be no *nitrate* run-off to cause *eutrophication* in rivers
- kinder to wildlife as there is no pesticide to kill insects and birds
- self-sustaining, as it produces more energy than it consumes
- less harmful to people as fewer chemicals are used

Cons:
- yields can drop considerably in the initial period when compared with conventional farming
- weeds can increase and need to be controlled by hand labor
- the produce is far more expensive when sold than conventional farming products.

organic matter: see *humus*.

organism: an individual living plant or animal. As well as plants, the organisms in a soil are the bacteria, the fungi and the animals that live in the soil such as earthworms and woodlice. They are more active and plentiful in warmer and more aerated soils. They are responsible for the decomposition of leaf litter, for the fixation of nitrogen in the air into *nitrates*, and for the binding together of individual soil particles to give a good structure. Burrowing animals also assist in the circulation of air and water through the soil.

Organization for Economic Cooperation and Development (OECD) was set up in 1961 and aims to coordinate international aid for *economically less developed countries (ELDC)* as well as providing a forum for discussion on economic growth and trade. Its members are mainly drawn from the *developed world*, but some of its most effective work has been in the collection and publication of worldwide economic and social data, standardized so that intercountry comparisons are possible.

Organization of Petroleum Exporting Countries (OPEC) is the name of the *cartel* that sets output *quotas* in order to control crude oil prices. The members of OPEC are countries in the Middle East, South America, Africa and Asia, but not the

USA, the former USSR or European producers such as the United Kingdom. In the 1970s, OPEC had great power when it controlled 90% of the world's supply of crude oil exports. Its influence, however, has been reduced, as the high price of oil allowed more expensive fields, such as Alaska and the North Sea, to be brought into production.

orogenesis refers to mountain building. Active belts of mountain building are formed either by *vulcanicity* or by the breaking and bending of the Earth's crust by *tectonic processes*, or a combination of both of these. Vulcanicity involves the accumulation of volcanic rock by the extrusion of molten material from within the Earth. Tectonic processes involve either the movement of continental crust toward oceanic crust, or the convergence of two continental crusts. In both cases intervening sediments are compressed, uplifted and folded to create mountains. The Himalayas are thought to have been formed in this way when the Indian subcontinent moved north to collide with the Eurasian plate and uplifted the sediments of the former Sea of Tethys.

orographic relates to processes caused by uplift over a mountain range. Orographic rainfall results when moist air is forced to rise over a mountain range. The rising air cools, the water vapor in the air condenses, clouds are formed, and eventually rain falls. Such rainfall is common on the western coast of Canada. As the air descends on the leeward side of the mountain, it warms *adiabatically*, condensation ceases, clouds dissipate and little rain falls. This is known as a rain-shadow.

out-of-town location: a site for an industry or shopping complex on the edge of an urban area. Here the land is cheaper, and there is more space for future expansion. Large parking facilities can be built. Such a site is frequently near a highway exit, making it easily accessible to places further away.

Larger out-of-town *malls*, such as the Mall of America in Bloomington, Minnesota, are also aimed at creating a day out for the family with an emphasis on shopping and other leisure activities.

outwash: the material deposited beyond the snout of a glacier by the *meltwater* streams that issue from it. Outwash consists of a gently sloping plain made up of sands and gravels. These particles are graded into different sizes, the larger ones being found near to the ice front and becoming progressively finer with increasing distance from it. The finest particles, rock flour, may be transported considerable distances before being deposited. Outwash deposits tend to be more rounded than glacial deposits due to having been rolled around in moving water. Outwash plains are characterized by multithread river channels, called *braiding*, and by small lakes, called *kettleholes*.

The deposits in an outwash plain are also graded vertically. In summer, the discharge of the meltwater streams is greater and they can transport larger particles across the plain. In colder periods discharge is much lower and the ability to transport is reduced. Consequently, a vertical stratification can be found with alternating layers of coarse and fine particles.

overgrazing is a major cause of *soil erosion* in areas with an arid or semiarid climate. The *nomadic* tribes who inhabit these areas measure their wealth in terms of the number of animals they own. In very dry periods of weather they concentrate their large herds around the few remaining waterholes, and the amount of grass available cannot cope with their needs. The vegetation cover becomes depleted, and the soil is exposed to erosion by either wind or water.

overland flow is the outcome of rainfall intensity on a slope being greater than the rate at which the water can infiltrate into the soil on that slope. A thin layer of water forms on the surface and begins to move downslope under gravity. Some of this water accumulates in small surface depressions, but these also overflow when full. Subsequent coalescence of this flow takes place and *rilling* occurs.

Overland flow of this type occurs in areas with high intensities of rainfall, devoid of vegetation. It can also occur in some cultivated fields, and where the soil has been compacted by vehicles or animals.

Saturated overland flow may occur in the lower parts of a slope. The ground nearest to the stream channel is saturated, and subsequent rainfall cannot seep into the soil. As the storm proceeds water flows over the surface and into the stream channel. The longer the storm continues, the larger the saturated area, and the greater the amount of run-off over the surface.

overpopulation exists when there are too many people in an area relative to the amount of resources, and to the level of technology locally available, to maintain a high standard of living. It implies that, with no change in the level of technology or natural resources, a reduction in a population would result in a rise in living standards. The absolute number or density of people need not be high if the level of technology or natural resources is low.

Overpopulation is characterized by low per capita incomes, high unemployment and underemployment, and outward migration.

overproduction is exemplified by the surpluses produced under the *Common Agricultural Policy* of the *European Union*. Farmers are given guaranteed prices for their produce and consequently have produced as much as possible. This has created excesses in a range of agricultural products such as the grain, butter, beef and yellow raisin "mountains" and the milk, wine and olive oil "lakes." These surpluses are stored by the governments of the European Union, or sold cheaply to developing countries.

overspill: people who are encouraged or forced by *redevelopment* schemes to move from a town or city to a new settlement beyond the limits of the original urban area.

oxbow lake: a horseshoe-shaped lake separated from an adjacent river. The water is stagnant, and in time the lake gradually silts up, becoming a crescent-shaped stretch of marsh. It is formed by the increasing sinuosity of a river *meander*. Erosion is greatest on the outer bank, and with deposition on the inner bank, the neck of the meander becomes progressively narrower. During times of higher discharge the river cuts through this neck, and the new cut eventually becomes the main channel. The former channel is sealed off by deposition.

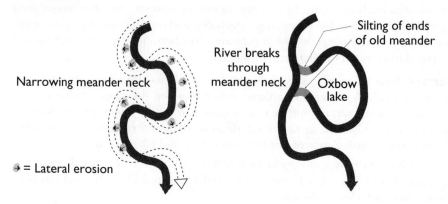

Oxbow lake

oxidation is a *weathering* process that involves the reaction between metal ions and atmospheric oxygen, or oxygenated water, to produce oxides or hydrated oxides respectively. An example is when iron II (ferrous) in mineral compounds is oxidized to form iron III (ferric). The structure of the original mineral compound is destroyed, as the oxidized mineral increases in volume. The rock containing the mineral is weakened, and becomes prone to further weathering.

Oxidation can be identified on rocks by the presence of a yellow or red-brown discoloration.

ozone depletion occurs when the rate at which the *ozone layer* is formed is less than the rate at which it is destroyed. In recent years, this depletion has manifested itself by the emergence of a "hole" in the ozone layer over Antarctica and the Arctic. Such "holes" are also beginning to appear over other parts of the world, albeit for limited lengths of time. The damage is believed to have been caused by the use of *chlorofluorocarbons (CFCs)*, which break down ozone. In both 1987 (the Montreal Protocol) and 1989 (the "Saving the Ozone Layer" Conference in London), a large number of countries signed a declaration to set in motion the reduction and phasing out of the production of CFCs.

ozone layer: a concentration of the gas ozone located in the *stratosphere* at an altitude of between 10–50 km above sea level. The ozone is formed by the interaction of solar ultraviolet radiation and oxygen in the atmosphere. This concentration shields the Earth from most of the harmful ultraviolet radiation from the sun. It is this radiation that causes an increase in the incidence of skin cancer and eye cataracts.

Ozone is also produced on the Earth's surface by car exhaust systems, and is damaging to both humans and plants.

P

Pacific Rim: the countries on the margins of the Pacific Ocean basin. These include Japan, Australasia, the western part of North America and the *newly industrialized countries* of Southeast Asia such as Hong Kong, Singapore, South Korea and Taiwan. Whereas at the beginning of the 20th century manufacturing was concentrated in North America and Western Europe, during the 1970s and 1980s plants were established by *transnationals* in the newly industrialized areas. The Pacific Rim now challenges the traditional manufacturing areas on the margins of the North Atlantic; this has been described as a *global shift.*

Pacific Ring of Fire: the name given to the *Pacific Rim* area where there is a *subduction* zone created by the oceanic plates sliding under those of the continent. The result is a line of volcanic activity that stretches through South America from

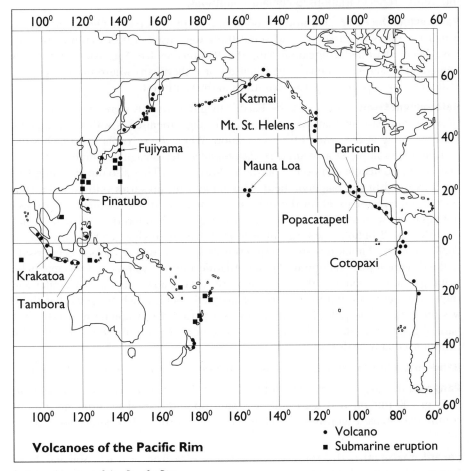

Active volcanoes of the Pacific Rim

southern Chile into Mexico and on into the Cascade Range of northeast USA. It continues through Alaska, the Aleutian Islands, Kamchatka, Japan, the Philippines, Indonesia, Papua New Guinea and on to New Zealand. This zone contains some of the world's most active and destructive volcanoes such as Mt. St. Helens, Tambora, Krakatoa (Indonesia), Pinatubo (Philippines).

paleoclimate literally means fossil climate, i.e. evidence in the present landscape that indicates that the climate was different when that evidence was laid down. This can include the evidence of the landforms that resulted from that climate; features that can only have been produced by the force of glaciers or ice sheets are found in areas with very different climates today. Fossil reef *corals*, which required warmer conditions for their growth, have been found in areas that do not support coral now. The evidence for reconstructing past climates are largely indirect – that is, the climatic conditions can be inferred from the effect the climate had on some other feature. These methods include *dendrochronology*, or tree ring analysis; *pollen analysis*; changes in coleoptera beetles; ice core and ocean core investigations of oxygen isotopes O^{16} and O^{18}. Each of these techniques works on the general principle that differences reflect variations in climatic conditions.

Paleolithic: the earliest cultural period of human prehistory, the Old Stone Age, in which roughly shaped stones and some flints were used. Various culture periods have been identified coinciding with the Pleistocene Ice Age; each of these is named after places in France (e.g. Chellean or Abbevillian) where there is greater evidence of occupation, e.g. cave paintings in the southwest.

paleomagnetism: evidence of the Earth's magnetic field is stored within rocks as "fossil magnetism" in that magnetic minerals align themselves with the magnetic field operating at the time of their formation. When *magma* cools as it reaches the surface, for example at a *constructive plate boundary*, ferromagnetic minerals behave like a compass needle and point toward the North Pole. They also dip at an angle to the horizontal in the magnetic field. The amount of dip varies with latitude from zero at the equator to 90° at the pole. In sedimentary rocks, magnetic grains also line up with the magnetic field. The Earth's magnetic field has reversed many times over the last 2 billion years; the time interval for these changes has varied from 20,000 to over 10 million years. These reversals are recorded in rocks on either side of the divergent plate margin in the mid-Atlantic and the symmetry in this magnetic striping provides evidence that the plates are moving apart. Rocks of a similar age show alignment to different patterns of magnetic field, which suggests that there was more than one pole. But, as at any one period the magnetic minerals must have aligned with one North Pole, this provides evidence indicating that it is the continents that have drifted.

pandemic refers to a disease that has affected people over a wide area of the world.

Pangaea: the large supercontinent proposed by Alfred *Wegener* as the original land mass of 200 million years ago that later split up and drifted to form the present continental areas. It consisted of *Laurasia* to the north and *Gondwanaland* to the south.

parent material is the rock or weathered *regolith* from which a soil has been formed. The mineral constituents and fragments in the soil are largely derived from the parent material, which exerts a particularly strong influence on poorly developed or *immature soils*. As time progresses, a soil generally becomes more independent of the parent material, and is more in balance with the climate and vegetation in the *biome*. Parent material is an important influence on soil texture, which reflects the mineral composition of the rock and the way it is weathered.

particle size analysis: a technique for measuring the percentage of each size of particle in a deposit. A sample of the deposit, for example a till, outwash material, or river alluvium, is collected in the field. A known weight of the sample is passed through a series of sieves with different mesh sizes. By weighing the contents of each sieve, and comparing this with the total weight, the percentage of the total deposit that lies within each grain class size can be calculated. The particle size distribution can be plotted on a *cumulative frequency graph*.

The main grain sizes are normally defined as follows:

Clays	<0.002 mm	(<2 μm)
Silts	0.002–0.06 mm	(2–60 μm)
Sands	0.06–2 mm	(60–2000 μm)
Gravel	2–64 mm	
Cobbles	64–256 mm	
Boulders	>256 mm	

Measurement in microns is often given in preference to millimeters, especially for the smaller grain sizes up to 2 mm. 1 micron is one thousandth of a millimeter, 1 μm = 0.001 mm; the equivalent grain sizes in microns are shown above.

particulates: see *atmospheric particulates*.

pastoral farming involves the deliberate rearing of livestock for meat or other products such as milk or wool. This can be *extensive* in operation as in the *commercial* ranching of beef cattle in east Texas or sheep in Australia; or it can be an *intensive* system such as dairying in the dense market areas of Western Europe. The activity ranges from the high technology of scientific breeding and capital investment associated with commercial systems to the low technology of nomadic herding groups who are very dependent on the environmental conditions within their marginal production areas.

paternoster lakes are produced by glacial scouring along a valley floor, which results in a series of lakes separated by depositional material or rock bars. As these lakes are linked together by streams, from above they resemble a string of beads; hence the term paternoster, a bead in a rosary.

patterned ground is a term that relates to the coarse and fine debris that accumulates on the surface, or just below, in *periglacial* areas. This includes features such as *stone circles*, garlands and stone stripes as well as *ice wedge* features. These are characterized by a degree of lateral sorting of debris. Daily fluctuations of temperature through freezing point and the annual *freeze-thaw* cycle results in the formation and decay of ground ice in the weathered surface layer. This produces frost heaving, which brings debris to the surface, forming a small mound. Ice pushes stones to the surface where they are sorted by gravity down the low angle slopes of the frost mounds. On steeper slopes the debris is elongated downslope to produce garlands and stripes.

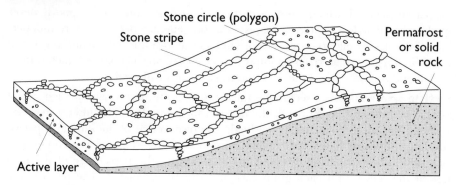

Patterned ground

peak flow is the maximum level of *discharge* in a storm *hydrograph* largely resulting from the arrival of fast *overland flow* into the channel system.

peak land value point: the point within the urban area where land values are highest. It usually occurs at a major intersection within the *central business district* where the *high-order* retail outlets tend to concentrate because of high pedestrian flow in this area.

peatland: an area that has a waterlogged, *anaerobic* environment favorable for the formation of peat. This could occur on flat uplands with impermeable subsoil or on valley floors that have a high water table. With waterlogging there is a lack of oxygen in the soil, breakdown of organic matter is very slow and material accumulates into thick layers of peat, which may produce extensive blanket *bogs*.

pediment: a low angle, rock cut surface at the base of a steep slope in a semiarid or arid area. They are broadly concave in profile and may be covered by a thin layer of alluvium, which is being transported from the steep slope above the pediment to a more gentle slope below the feature. The presence of alluvium suggests that water has been involved in the formation. However, their origin has been the focus of considerable discussion. Some researchers view the pediment as a basal slope, which is the remnant of a mountain front undergoing parallel retreat, its slope transporting weathered sediment away from the steep front to the flat area beyond the pediment. Others suggest that the feature is the result of undercutting by running water in the form of sheet floods causing it to recede. The effect of sheet floods on semi-arid landforms is controversial; some argue that sheet floods do not produce this feature, but only develop as sheet floods because the feature already exists. It may be that both theories play a part in the formation.

pedogenesis: the formation of soils. The factors that influence soil development are as follows:

- climate – this affects the rate of weathering of the *parent material* and *regolith*. The balance between precipitation and evaporation determines the *soil moisture* budget, which influences the extent of downward *leaching* or upward *capillary movement* of minerals
- organic material and organisms – plants, bacteria and animals interact in the *nutrient cycle*, returning minerals to the soil store to be used again.

Organic matter provides material for *humus* within the soil, which contributes to the clay-humus complex

- relief – as altitude increases rainfall total tends to increase, and temperature and the length of the growing season decrease, influencing the nature of the soil. Angle of slope affects drainage, run-off and soil depth. On steep slopes, faster *throughflow* and surface run-off increases downslope movement of soil particles, which will retard soil development. Upper slopes are shedding sites, i.e. water is draining away, which may promote leaching; lower angle zones at the base of slopes are receiving sites, which may experience waterlogging and *anaerobic* conditions

- *parent material* – this influences the mineral composition of the weathered material forming the basis of the soil and the degree of permeability of the subsoil, which will affect soil drainage. Depth and *soil texture* will be influenced by the ease of weathering and the mineral composition of the parent material

- time – young, poorly developed soils that are *immature* tend to be strongly influenced by their parent material, but mature soils that have become independent of the parent material are in balance with the *ecosystem* and tend to reflect the climate and vegetation more closely.

peds: soil units produced by the grouping together of sand, silt or clay particles. Peds are bound together by the gums and mucilages formed during bacterial breakdown to create *soil structure*. Peds are the "building blocks" of the soil and the spaces between peds house microfauna, which are important to soil formation. Soil ped arrangements (structure) fall into five categories: structureless, platy, crumb, blocky and prismatic.

Pelean eruption: a violent form of volcanic eruption, which is accompanied by an explosion of gas, ash, and pyroclasts in the form of an incandescent cloud, *nuée ardente*, which travels at great speed down the flanks of the volcano. It is named after Mt. Pelée on the island of Martinique in the Caribbean, which erupted violently in 1902.

per capita means per head and is often used when making comparisons between regions or countries that have different base populations, e.g. when quoting an individual measure such as income, *GNP*, or calorific intake per capita.

perception is the way in which an individual or a group views a particular environment or situation. For example, different groups will view and interpret the attractions and hazards of a landscape in very different ways according to their perception. Perception is the process of evaluating and storing information that is received; this information may be part of the cultural background or tradition in society, or it may be provided by the media in the form of TV documentaries, newspapers, news bulletins and films. Information may be transferred verbally from one person to another. Some, or all, of these sources could influence the image an individual has of a particular place or environment. This has led, for example, to the popular misconception of the West as a place inhabited by people wearing stetsons and lassoing cattle.

Whereas many *models* are based on the concept of *economic man* – the *optimizer* who has perfect knowledge – behavioral approaches accept that decision-makers operate within "bounded knowledge," i.e. incomplete information. A decision-maker's perception of an environment can influence the decision taken.

percolation: the downward vertical movement of water within a soil. The water then enters the *groundwater* store. The rate of percolation depends on the size of the pores through which the water travels. Sandy soils have high rates of percolation because of the large voids between each particle, whereas clay-enriched soils have low rates due to the very small voids between clay particles.

perennial: lasting through all seasons of a year, or existing for several years. Perennial may be applied to:

- plants that live for several seasons or years
- irrigation schemes where the water is stored behind a dam and released at regular intervals to maintain a constant water supply throughout the year.

periglacial literally means on the fringe of, or near to, an *ice sheet* or *glacier*. It is also used to describe any area that has, or has had, a very cold climate. This includes high mountainous areas, the *tundra* areas of northern Canada and Northern Europe, as well as most of southern England during the Pleistocene Ice Ages.

periglacial landform: a feature created largely through the actions of *periglacial processes*. (See *active layer, blockfield, ice lens, ice wedge, permafrost, patterned ground, pingo, scree, stone circles, thermokarst*.)

periglacial processes are all caused by the ground freezing hard and often for considerable periods of time, yet also thawing for short periods of time. They take place in *periglacial* environments.

The processes include:

- *solifluction* (see also *active layer* and *permafrost*)
- frost heave (see *ice lens, stone circles* and *patterned ground*)
- thermal contraction (see *ice wedge*)
- frost weathering (see *blockfield* and *scree*)
- thermal subsidence (see *thermokarst*).

periphery: an area of low or declining economic development, associated with the models of Gunnar *Myrdal*, Albert *Hirschman* and John *Friedmann*. It is usually applied to a region within a country suffering from high unemployment, high rates of selective outmigration, low personal incomes, low living standards and high rates of crime. The periphery is often in an unfavorable geographical location, with a poor and/or deteriorating resource base.

Periphery can also refer to those parts of a country that do not constitute the *core* of that country.

permafrost is perennially frozen ground. It lies beneath approximately one-fifth of the world's land surface, with extensive areas in Russia, northern Canada and Alaska. It also exists beneath the sea in the Beaufort Sea, and at high altitudes in the Rockies and Central Asia. In northern Russia, the permafrost extends to depths of over 1000 m.

In some areas, there exists above the permafrost an *active layer* – soil that thaws out seasonally.

There are several types of permafrost:

- "wet" permafrost – where all of the pore spaces and voids are filled with ice
- "dry" permafrost – where the soil and rock are unsaturated
- continuous permafrost – where it is present in all localities
- discontinuous permafrost – where there are some small scattered unfrozen areas
- sporadic permafrost – where there are only a few patches of permafrost, for example on a poleward-facing slope, in an otherwise unfrozen area.

The development of permafrost is due to:

- very cold temperatures all year around, the mean annual temperature being below −5°C
- low precipitation totals, so that snow does not insulate the ground from the cold
- a limited vegetation cover.

permeable: a description of a substance that allows water to pass through it. Permeability can be divided into two types:

- *porosity*
- perviousness – this occurs when rocks have *joints* or fissures along which water can flow. Examples include *carboniferous limestone*, which has its joints and bedding planes widened by solution, and granite, which has joints because of the manner in which it has been formed. (See *impermeable*.)

pesticide: a chemical applied to crops to control pests and disease. Some people have expressed concern at the rate at which the use of pesticides is increasing. Instructions for usage are not always followed correctly, safety regulations are not always conformed to, and equipment is not always properly maintained. There have been several claims that pesticides have damaged human health in some farming areas, and may be the cause of birth abnormalities. Pesticides have affected "innocent" wildlife in many areas, killing birds, bees and other fauna.

pH indicates acidity by measuring the concentration of hydrogen ions. This is represented on a scale from 0 to 14, which is logarithmic (pH3 is 10 times more acidic than pH4 and 100 times more acidic than pH5). This is used particularly when describing soils. Normal rainfall has a pH of between 5 and 6.

phosphates are deposits that are mined for use as *fertilizers*. As rock phosphate is in limited supply, large amounts of fish can be converted to phosphorus-rich fertizers. Another source is where there are masses of fish-eating birds: these birds excrete matter rich in phosphorus, which is then mined if the deposits have built up to the extent that they are commercially exploitable. Such deposits are known as guano.

photochemical smog: a form of *smog* that occurs in large cities and can be dangerous to health. When exhaust fumes and factory emissions are trapped by *inversions* of temperature, they can combine with sunlight to produce ozone. Such smogs can cause breathing difficulties, vomiting, eye irritation and a general lethargy. In some photochemical smogs the presence of substances containing nitrogen could increase the risk of cancer.

photosynthesis: the green pigment chlorophyll, present in green plants, is able to capture light energy and convert it into food energy by manufacturing carbohydrates (energy storing chemicals) using elements such as oxygen, carbon and hydrogen. Photosynthesis, therefore, allows the plant to grow and increase its *biomass*.

FORMULA $6CO_2 + 6H_2O + \text{solar energy} \rightarrow C_6H_{12}O_6 + 6HO$

$\qquad\qquad$ Carbon dioxide $\quad$ water $\qquad\qquad\qquad\qquad$ glucose $\quad$ oxygen (released)

photovoltaic cells contain semiconductor crystals of silicon (or gallium arsenide) that, when exposed to the sun, generate electricity, which can be stored in batteries. They are used in the *Third World* for small-scale local electricity production, but a wider application has run into problems of scale, cost and maintenance.

physical weathering involves the breakdown of rocks into smaller fragments through mechanical processes such as expansion and contraction due mainly to temperature change. Two such types of physical breakdown are *freeze-thaw* weathering and insolation weathering or thermal fracturing. In addition salt weathering, caused by the growth of salt crystals, and *pressure release* or dilatation can also bring about weathering without temperature changes.

The effect of these weathering processes is to widen joints and cracks, which allow deeper penetration of weathering action, and the surface may become littered with fragments that are then detached from a rock outcrop; surfaces may be buried under a layer of weathered debris, the *regolith*. The products of weathering vary according to the constituent minerals in the rock. Breakdown may produce large blocks, block disintegration, or smaller fragments that reflect granular disintegration. Some researchers classify biological, or biotic, weathering as a separate form of breakdown, but others regard some of the biological processes as physical in operation; tree and plant roots penetrate joints and prise the rock apart in a process that is similar to joint widening by freeze-thaw action.

phytoplankton are very small plants that float near the surface of seas and oceans. Like land-based plants, they produce food by *photosynthesis*, obtaining their carbon from atmospheric *carbon dioxide* dissolved in ocean water and fueled by sunlight that only penetrates the surface layers of the oceans. Phytoplankton form part of the base of the marine *food chain*. In places, they often form dense blooms, which may have a density of 200,000 individual plants per cubic meter of sea water.

Pick Your Own: a method of marketing that cuts out the middlemen in the *agricultural chain*, such as *wholesalers*. Farmers are therefore able to increase their profits, and customers benefit from very fresh produce at prices below the market level.

pie chart (graph): a representational technique consisting of a form of *proportional circle* in which the area of the circle is divided into segments. These segments represent a percentage of the whole figure.

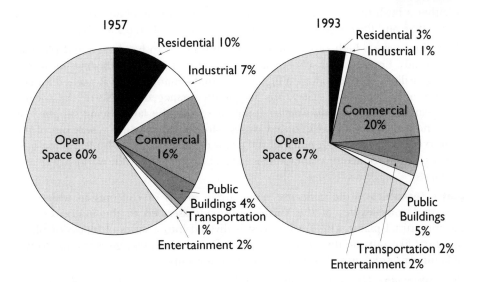

Pie charts: example of land use in Our Town, USA

piedmont: the area at the foot of the mountains, as used in the term piedmont glaciers. These are formed when valley glaciers flowing from the mountains extend onto lowland areas, spread out and merge. A well-known example of a piedmont glacier is the Malaspina Glacier in Alaska.

pingo is a dome-shaped isolated hill that stands out on the flattish *tundra* plain. They are formed by the freezing and consequent expansion of water within the upper layers of the ground.

pioneer: a plant or a community of plants that are the first to colonize an area. These are generally simple and hardy plants that will ultimately improve the conditions allowing other plants to grow. Under different conditions, different pioneers will colonize the area. On bare rock (*lithosere*), for example, the colonizers will be lichens and mosses, but on a sandy area (*psammosere*) the pioneers will include sea couch, lyme and marram grass.

plagioclimax: when human interference has permanently arrested and altered the natural vegetation so that the climatic *climax vegetation* has not occurred, the resulting community is known as plagioclimax.

plagiosere is a plant *succession* that has been shaped by human action.

planning blight is where uncertainty and delay in the development of a particular urban plan can cause an area to deteriorate as people refuse to invest in property that might be demolished. If the plan is delayed for some time, residents wishing to move suffer, as they cannot find buyers even for their devalued property.

plantations are farms with large-scale agricultural activity aimed at producing cash crops of high value. They were originally developed in tropical areas by North American and European merchants. Large areas were cleared, and a single crop of either a bush or tree was planted. They are an example of an extreme form of *monoculture*, and the crops include sugar cane, coffee, rubber, tea and bananas.

Plantations have a high capital input. The land needs to be cleared and planted; estate roads, housing, and schools need to be built, together with some local processing facilities. They also have a high labor input – manual labor is needed for cultivation and harvesting. They are frequently managed by North American/European/Pacific Rim *transnational* companies, with the labor being supplied by indigenous people who work for low wages. Most of the produce is sent out of the country of production, and much of the profit of the enterprise also leaves the host country. The produce of plantations is subject to fluctuations in world prices and demands.

plate tectonics is the collective name for a group of concepts in which the structural complexities of the Earth's crust are ascribed to the interactions of moving crustal plates. The Earth's crust is divided into a series of blocks or plates that float like rafts on the underlying *mantle*. Their movement gives rise to a series of landforms, submarine features and hazardous events.

(See *conservative plate margin, constructive plate margin, crust, destructive plate margin, earthquake, folding, island arc, mid-oceanic ridge, ocean trench, sial, sima, volcano*.)

playa: a shallow, *ephemeral* and saline lake found in desert or semidesert areas of the world. It is formed after rare rainstorms, after which the water evaporates rapidly leaving a very flat area of clay, silt or salt. Some larger playas, for example the Dead Sea in Israel and the Great Salt Lake in Utah, have not dried out completely. However, the latter is only a vestige of what was once the much larger Lake Bonneville, the former bed of which is used for attempts at the world land speed record.

Pleistocene: a geological time period stretching from 2 million years before present (BP) to 10,000 years BP. It was characterized by a series of alternating cold phases (glacials) and warm phases (*interglacials*) known collectively as the *Ice Ages*.

plucking: a process of glacial erosion by which a *glacier* freezes around a rock on a valley side or bottom, and subsequent movement of the ice causes the rock to be pulled away with it. It is believed that this can only be effective when the bedrock has been well-weathered, or is well-jointed, prior to the glacier passing over it. As the process goes on, some of the underlying joints in the rock may also open up as the weight of the overburdening rock is removed. This process of *pressure release* enables plucking to continue to operate.

pluvial: a period during the *Pleistocene* when wetter conditions existed in those areas of the world that are currently arid. There is much evidence of more water being available in desert areas in the past:

- the beds of huge former lakes that can be found in the southwest USA, such as Lake Bonneville
- expanses of fossil soils of a humid type, including laterites
- vast river systems in the central Sahara that are now dry and inactive
- plant and pollen remains (oak and cedar) and animal fossils (antelope and rhino) in the Sahara
- evidence of former human occupation of Neolithic culture.

podsol: a type of soil associated with Northwest European and Siberian type climates, and *coniferous woodland* or *heathland* vegetation. In these types of environment, evapo-transpiration rates are less than precipitation, causing water to move downward through the soil. Due to the cool and wet climate the decay of the needle-shaped leaves is very slow, and a thick acidic form of peaty humus (*mor*) is created on the surface.

The downward percolation of acid water through the soil, particularly after snowmelt, causes the rapid *leaching* of soluble bases and the movement of clays (*lessivage*), and the *eluviation* of iron and aluminum sesquioxides (*podsolization*). The result is that the upper layers of the soil become depleted of these substances and bleached of color, leaving a sandy ash-gray horizon. However, they are redeposited lower down the profile. At first, a very thin layer of redeposited iron, called an *iron pan*, may be found in some podsols. Beneath this, a darker horizon consisting of redeposited humus, and a clay-enriched orange-brown horizon are found.

The cold climate also discourages organisms, and there are few earthworms to mix the soil horizons. Horizon boundaries are therefore sharply defined. Podsols are not naturally fertile because of their acidity. They can be improved by deep plowing to mix the upper and lower horizons, and by the frequent application of lime and fertilizers.

podsolization is an intense form of *leaching* that operates in cool climates where the rate of precipitation is significantly greater than the rate of evapotranspiration. The vegetation of such areas is *coniferous woodland* and *heathland*. These produce a highly acidic form of humus – *mor*. Rainwater passing through this becomes highly acidic, so much so that it is capable of breaking down clay particles and mobilizing the *sesquioxides* or iron and aluminum. These are removed from the upper part of the soil, leaving a bleached ash-gray horizon. They are redeposited lower down the soil, in some cases forming a thin *iron pan*. The resultant soil of this process is a *podsol.*

point bar: a gently sloping bank on the inside of a *meander* composed of sands and gravels. The water in the river is moving more slowly here and this results in deposition. Initially coarser particles are deposited at a point just downstream of the steepest part of the inside bend.

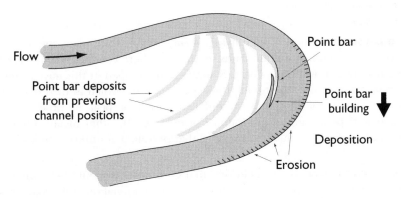

A point bar

As this grows, finer sediments are deposited in the shallower and calmer water between it and the inside bend. Vegetation may encroach, trapping more sediment.

polar: used to describe natural phenomena that owe their origin to being at or near to either of the poles of the Earth. Some examples include:

- polar air masses – associated in the northern hemisphere with northern Canada, northern Russia and the Arctic Ocean
- the polar front – the zone where tropical and polar air masses collide in the high and mid-latitudes of each hemisphere. In the northern hemisphere this is in the North Atlantic ocean at approximately 50–60°N (see *fronts*)
- the polar cell – a vertical circulation of air in the high latitudes. This involves the upward movement of warm air at the polar front, the horizontal movement of this air at high altitude toward the poles, the subsidence of the air at the poles, followed by a horizontal movement along the surface of the Earth back to the polar front
- polar climates – the equatorward limit of which is generally taken as the line where the mean temperatures of the warmest month is no more than 10°C.

polarization is the uneven development that often takes place in *Third World* countries as a result of the emergence of a *core*. The core area attracts investment particularly in *infrastructure*, often to service *transnational companies*. This is often at the expense of the *periphery*, which loses resources, investment and young people to the core.

The term can also be used to express the accentuation of a difference between two things or groups, or alternatively, the process of division into two groups representing the extremes of opinion, wealth, or the like.

political system: the means by which government and/or administration can take place in a country. A wide range of systems exist around the world, and have existed in the past. Types of systems, which are not always mutually exclusive, include:

- totalitarianism – ruled by one governing party or individual. No rival loyalties or parties are permitted
- democracy – government by the people, either directly or through elected representatives
- *command economies*
- *capitalism.*

pollen analysis: a method of identifying *climatic change*. Each plant species has a distinctive pollen grain. If a grain lands in an area where conditions are *anaerobic*, such as a peat bog, it will resist decay. Several grains trapped in this way are therefore representative of the vegetation growing at that time, which in turn is a reflection of the climate of that time. If differing layers of peat in which varying combinations of pollen grains are found can be dated by other means, then it is possible to ascertain the climatic conditions that prevailed at different times in the past.

polluter pays: the increasingly held view that governments should force the originators of *pollution* to pay the costs of removing the contamination and making good the damage that has been caused.

pollution means contamination of some kind. It is usually referred to in the context of the environment, where the pollution can take many forms – air, water, soil, noise, visual and many others.

pools and riffles: on the *meanders* of a river, the pools are the areas of deeper water, whereas the riffles are the shallower parts. The pool represents the area where the energy of a river builds up due to a reduction in friction, but across the riffle energy is dissipated. Here, a higher proportion of the river's energy is required to overcome friction so that deposition occurs. Pools and riffles tend to be regularly spaced along the course of a meandering river.

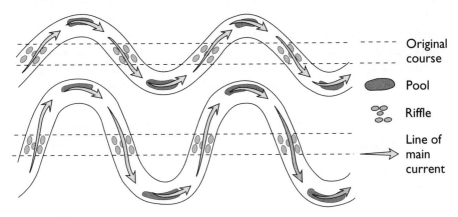

Pools and riffles

population: apart from the straightforward meaning of the term in reference to people, population is also a statistical term meaning the whole from which a *sample* will be chosen for a research exercise. This could be people, but it could equally be pebbles on a beach, farms in an area or houses in a town.

population change: the population of an area increases or decreases according to the *birth* and *death rates* as well as the movement of people in and out of the region (*migration*).

population density: the number of people per unit area. The density of population is obtained by dividing the total population of a country (or any region) by the total area of that country (or region). The density of population for a country can be misleading as it does not show variations between densely populated regions and those areas that are almost unpopulated. Population densities are often shown on *choropleth* maps.

population distribution describes the way that people are located within an area. When plotted on a map, this can reveal those areas that have a high population and those where few people live, which maps of *population density* fail to show. One of the ways of plotting this information is by means of a dot map.

population policies: in the 1970s many *Third World* countries began to formulate policies with regard to the growth of population. Most countries now have policies with regard to:

- the growth of the population
- the level of *fertility*
- levels of mortality
- the spatial distribution of population within the country
- international *migration.*

In most of the countries the policies are directed toward attempting to control the high population growth rate. Programs include *birth control,* abortion, education, sterilization and social and economic incentives. Among the best-known of these policies is the one-child policy in China. Here, draconian measures are brought to bear upon families exceeding the quota, such as increased taxation, loss of housing subsidies, etc. The *United Nations Fund for Population Activities (UNFPA)* holds conferences on population every ten years, the latest taking place in *Cairo* in 1994.

population pyramid: the *population structure* of a country is best shown by a population or age-sex pyramid. The diagram is known as a pyramid, although the structure of the population of many countries does not take on this shape.

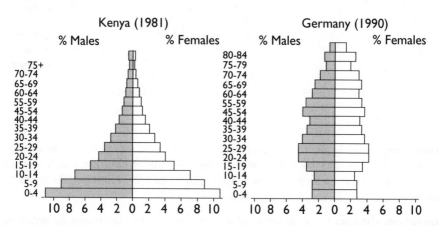

Examples of population pyramids

population structure is the make-up of the population of a country (or area). The most studied form of structure is that of age and sex and is represented by a *population pyramid.* The age, sex and life expectancy of a population has implications for the country's future economic and social development. Other structures that can be studied include race, language, religion, family size, etc.

porosity refers to the amount of area between the particles of a rock. Such areas are called pore spaces and the size and alignment of them determines how much water can be stored or can pass through the rock. Porosity is sometimes expressed as the percentage of the total rock taken up by pore spaces, e.g. sandstones range from 5 to 15%, loose sand and gravels can reach 45% and clays up to 50%. Saturation occurs when all the pore spaces are full of water.

positive correlation: a *correlation* where an increase in the value of one *variable* is matched by an increase in the value of the other. For example, the number of automobiles per head of population will show a positive correlation with *GNP.*

positive feedback: this occurs within a *system* where a change causes a snowball effect, continuing or even accelerating the original change. For example, intense heating in the tropics causes *convective* uplift over the oceans. This draws in moist air and the system is continually fed through the release of *latent heat* from the massive *condensation* and *precipitation* that occurs within the rising air. A small disturbance can then rapidly grow into a *tropical* storm and perhaps even into a *hurricane.*

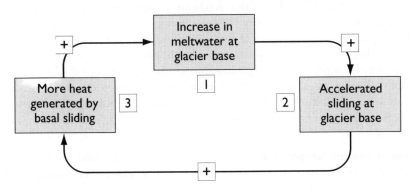

Positive feedback

positive skew is a bias within a *distribution* toward low values.

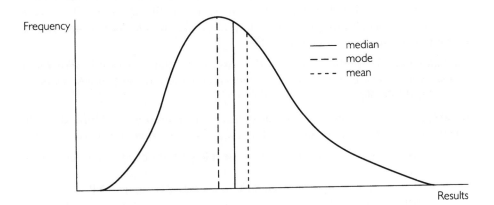

Positive skew

postglacial: the period since the retreat of the last *Pleistocene* ice advance. In North America and Europe the last ice sheets retreated about 10,000 years *BP,* and there followed a series of climatic changes with a rise in temperature to the climatic optimum during the Atlantic period, which ended about 5000 years BP. Since then the climate has become cooler. *Pollen analysis* and studies of oxygen isotope ratios in *ice and sea cores* has enabled the postglacial to be divided into a number of climatic periods: Pre-Boreal (10,000–9000 BP); *Boreal* (9000–7500 BP); Atlantic (7500–5000 BP); Sub-Boreal (5000–2500 BP); Sub-Atlantic (2500 BP to present).

These changes are closely linked to the general *vegetation succession* that developed after the Ice Age. In the Pre-Boreal the climate was cold and dry and species such as the birch and pine were common. Temperatures modified slightly in the Boreal, which was dry with cool winters but warmer summers. The amount of pine and birch gradually decreased and species like the elm, hazel and oak, trees that prefer warmer conditions, became more significant. The Atlantic period was warm and moist, perhaps 3°C warmer than today, and this represents the *climatic climax* period for the mixed *deciduous woodland* with oak, elm and hazel being joined by more temperature sensitive species such as lime and alder, which tend to appear in the later stages of forest development. The Sub-Boreal was cooler and drier; there was some reappearance of birch, but generally a stable vegetation. Human activity begins to have a greater impact during this period (the *Bronze Age*) and pollen from grasses, weeds and plantains appear in the pollen spectrum, which suggests some small-scale clearance and cultivation. The Sub-Atlantic is a warmer and wetter climate; vegetation changes became increasingly affected by human activity.

potential evapotranspiration: the amount of water that could be evaporated or transpired from an area given sufficient water available. Potential evapotranspiration is directly related to prevailing temperature conditions.

The relationship between *precipitation* and potential evapotranspiration for an area is represented by a *soil moisture graph.*

potholes are cylindrical holes drilled into the bed of a river by turbulent high velocity water loaded with pebbles. Vertical eddies in the water may be strong enough for the pebbles to grind a hole into the rock. Potholes vary in size from a few centimeters to several meters in width.

The name potholes is also applied to the passageways found beneath *carboniferous limestone* areas. These are created by the combination of sink holes, caverns, widened bedding planes and joints, and springs.

poverty: an abstract notion of extreme deficiency or inferiority, which is extremely difficult to quantify. It can be measured by a wide range of indices such as:

- food supplies – some 800 million people go without sufficient food every day
- per capita income – illustrated by an average per capita GNP of less than $400 per year
- access to clean drinking water and other sanitation services
- access to primary health care facilities
- adult *literacy* rates – nearly 900 million people are still unable to read or write
- *quality of life* indices.

Poverty in the *economically less developed countries (ELDC)* of the world poses one of the greatest threats to the environment. This is because, in extreme poverty, a person's immediate need is to survive, and not to have a concern for the longer-term future.

precipitation: water in any form that falls from the atmosphere to the surface of the Earth. It includes *rainfall, snow, sleet,* and hail.

precipitation intensity refers to the rate at which *rainfall* falls within a given period of time. Low intensities of rainfall, sometimes called drizzle, tend to have longer periods of duration. High intensity rainfall periods tend to have a shorter duration, as in the case of a *thunderstorm*. Intensity can also vary spatially. The center of a storm tends to have high intensity rainfall, whereas the edges of the storm have lower intensities.

precipitation processes: the mechanisms by which tiny cloud droplets become significantly larger rain droplets, and then fall to the ground surface. Research has identified two main processes:

- the *Bergeron–Findeisen theory*
- the coalescence theory. Unlike the above, this accounts for the occurrence of rain from clouds within which freezing does not take place, as in the tropical areas of the world. It assumes air turbulence within the cloud, which causes masses of tiny cloud droplets to be swept up and down within the cloud. These rising and falling air movements cause the droplets to collide with each other, and to grow in size. Eventually their weight becomes so great that they can no longer be supported by the rising air currents, and they fall as rain.

prediction: a form of hazard management whereby researchers try to forecast the occurrence of a hazardous event. Successful prediction enables people to be forewarned about an event, and therefore to make preparations to reduce the impact of it.

preservation: the maintenance of a landscape or building such that its current state is as close as possible to its original condition. Such features may be preserved as sources of tourist income and/or to provide an educational resource.

pressure (atmospheric): the weight exerted by the gases that constitute the atmosphere on any surface exposed to it. The atmosphere consists of a mixture of gases, which are confined by the ground or sea from below, and by the force of gravity from above, thus preventing them from escaping. The layers of the atmosphere closest to the ground surface have the greatest weight acting upon them, and so pressure is greatest here. Consequently, pressure decreases with height. Atmospheric pressure also varies horizontally, being a direct function of temperature. When temperatures rise, air expands and rises by convection, and pressure decreases. Conversely, when temperatures fall, air contracts and becomes more dense, causing an increase in pressure.

Atmospheric pressure is measured in millibars, and the average sea level value is 1013 millibars. Points of equal atmospheric pressure are shown on a weather map by lines called isobars.

pressure gradient is the rate at which barometric pressure changes spatially between high and low pressure areas. It is indicated by the spacing of the isobars on a synoptic chart: the closer the spacing the steeper the gradient. The speed of winds is influenced by the steepness of the pressure gradient. A simple analogy can be made with the link between slope gradient and the spacing of contour lines.

pressure melting point is the temperature at which ice under pressure will melt. Normal melting point is 0°C at the surface of a glacier or ice sheet, but at depth the increased pressure marginally lowers the melting point. When ice moving down a valley meets an obstacle, the basal ice will be subjected to additional pressure. If it is close to melting point, as it is likely to be in a temperate or warm *glacier*, then the additional pressure applied to the ice will cause melting. Temperate glaciers are close to pressure melting point throughout their thickness, but polar glaciers are normally well below 0°C and the pressure melting point.

pressure release or dilatation occurs when rocks, such as granite formed within the Earth's crust, are gradually exposed by denudation and the removal of overlying strata. These rocks, which were subjected to great pressure when buried under other strata, expand as the pressure reduces. This is known as sheeting, where layers of rock separate and form a system of jointing generally parallel to the surface. Pressure release, or "unloading," can also result from removal of rock by glacial action as the density of ice is lower than the density of the underlying rock.

prevailing wind: the most frequently occurring wind direction at a given location.

primacy occurs when the largest city dominates the city size distribution of a country. If a country followed the prediction of the *rank size rule*, the second city would be half the population of the first city. If the population of the first city is greater than twice the size of the second city then this can be described as a primate distribution. With primacy, the largest city is usually the capital and leading political, economic and cultural center in the country. It has a disproportionate share of activity, and this extreme centralization may attract more population through *migration*. It is difficult to generalize about the factors that encourage primacy; both economically advanced and less developed countries display primacy; former colonies, such as Sri Lanka, may display primacy as development was focused on one point of entry, although others such as Brazil display a binary pattern with two dominant cities of roughly equal size.

primary data is information that is collected through a *personal field investigation*, or material derived from other sources that has not been processed. This would include material such as that contained in a *census*, telephone directories, trade directories, etc.

primary sector: economic activity directly concerned with the extraction of natural resources; this includes mining and quarrying, agriculture, fishing, forestry, hunting. These are sometimes referred to as extractive industries.

primary succession: a vegetation succession that takes place on a surface where no soil or vegetation has formerly existed. Examples of "new" surfaces include natural surfaces such as sand dunes, lava flows, tidal marshes, landslips, outwash plains and land exposed by the retreat of ice, as well as surfaces created by human activity such as abandoned quarries, derelict land resulting from urban clearance, mine waste and spoil heaps. Primary successions can be divided into two broad types: *xeroseres*, which develop in dry conditions; and *hydroseres*, which are initiated in standing water. Xeroseres can be either *lithoseres*, which develop on bare rock surfaces, classed as dry because initially there is no soil to retain moisture, or *psammoseres* (sand dune systems), which are dry because of the extreme permeability of the surface.

Hydroseres are subdivided into hydrosere successions, where they develop in fresh water environments such as a lake or pond, and *haloseres*, where the succession occurs in salt water, such as a tidal marsh or estuarine mud flats. Although each succession follows its own stages of development, they all terminate in the *climax community* for that climatic type.

primeur crop: agricultural produce marketed out of season or early in the season. It particularly applies to fruits, vegetables and flowers that have a "year-around market" in advanced urban areas. This demand is met by bringing in crops from other areas where warmer conditions in winter or early spring allow crops to be harvested and marketed earlier than in the market area. Florida, with its mild winters, markets "early" produce in the urban area of the Atlantic coast from Washington through to Boston.

primogeniture is a form of land inheritance in which the estate is passed on without subdivision to the eldest son (male primogeniture) or to the eldest surviving child. The advantage of this form of inheritance over *gavelkind* is that the lands of the estate do not become fragmented.

prisere: a succession of plant community stages, or seres, that make up a *primary succession*.

private sector is that part of the economy operated by firms that are owned by shareholders or private individuals. In economies such as that of the USA it is the dominant sector; the remainder is called the *public sector*.

privatization (denationalization) occurs when firms or even whole industries are sold to the *private sector* after having been state-run. At its simplest, privatization can involve *contracting out* services within a local community (refuse collection, parks and gardens) or the *deregulation* of services. At its most extreme level, it involves the change of large state-run industries into public companies.

Arguments in favor of privatization:

- state control is not the best way to run industries as efficiency can only come about through the market mechanism
- many state industries are *monopolies* that act against the public interest
- selling shares widens the share ownership base within the country
- privatization provides government revenue that can keep taxes down
- there should be better industrial relations and *productivity* within the industry.

Critics have claimed that:

- in some cases the state monopoly has given way to privatized monopoly as there is really only room for one producer in the market
- this policy is akin to "selling off the family silver."

production chain: the entire sequence of activities required to turn *raw materials* into a finished product for the consumer. The chain includes *primary, secondary* and *tertiary sector* activities with the latter involved at every stage. A highly simplified example for beer production would be:

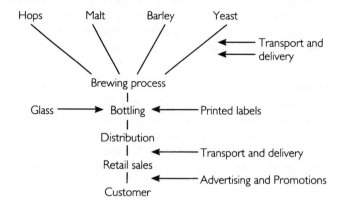

Production chain for beer

production line: the arrangement of a *flow production* system so that parts move systematically from one stage to the next.

productivity is a measurement of the efficiency with which a firm turns production *inputs* into output. The most common measure is labor productivity, i.e. output per worker. This is important because output per worker has a direct effect on labor costs per unit of production. The higher the productivity, the lower the labor costs per unit of production, as in the following example:

Worked example: productivity and costs for widget manufacture

	Weekly salary	Productivity (output per worker)	Labor cost per unit
Best UK firm	$500	25.0	$20
Worst UK firm	$500	12.5	$40
Average Japanese firm	$600	40.0	$15

Differing levels of productivity (efficiency) are the main single explanation for variations in industrial performance and levels of national wealth. Rising productivity can cause job losses, if demand does not rise as fast as productivity gains. Yet despite this threat, high productivity remains vital to the competitiveness of every company and country.

proportional symbols: representational techniques, where a symbol is drawn proportional to the value that it represents. This can be in the form of circles, spheres, bars, squares and cubes. Perhaps the most widely used is the proportional circle, where it is often found as a divided diagram, the divisions representing the components that go to make up the whole. Proportional divided circles are perhaps better known as pie graphs or *pie charts.*

protectionism describes policies of erecting barriers to trade such as *quotas, tariffs* and nontariff barriers. Because of the benefits that stem from trade, protectionism is banned under *GATT* agreements. The worst period of protectionism followed the

Great Depression of 1929, as many countries put up such barriers in the hope of maintaining levels of employment.

psammosere is a *succession* of plants that develops on sand, best seen on coastal sand dunes.

public corporation: the technical term for a *nationalized industry*, i.e. an enterprise that is owned by the state, but offers a product for sale to *public* and *private sector* customers.

public sector: the organizations and activities that are owned and/or funded by national or local government. These include *public corporations* (*nationalized industries*), public services and municipal services (those run by local governments).

push and pull factors encourage people to migrate, being the observations that are negative about the area in which the person is presently living (push) against the perceived better conditions in the place to which the migrant wishes to go (pull). A good example would be the factors that encourage rural-urban migration in the *Third World*:

Push: • imbalance between resources and population, causing *overpopulation*, *under-* and unemployment
 • rural poverty and low wages
 • physical environment often unfavorable for agriculture
 • modernization of farming leading to unemployment
 • lack of social services such as education, health and housing
 • break up of traditional communities through selective outmigration
 • *hazards*, including flooding, *earthquakes* and *volcanic eruptions*

Pull: • *perception* of job opportunities
 • higher salaries in urban areas
 • better provision of housing, education and health facilities
 • the attraction of the cultural and recreational activities of a large urban area.

pyramidal peak is formed when three or more adjacent *cirques (corries)* develop on the side of a mountain leaving a very sharp mountain peak with steep sides and *arêtes* radiating from the central peak – for example the Matterhorn in the Alps (for diagram see *arêtes*).

pyroclastics are materials that have been blown into the atmosphere by *volcanic* activity. These include cinders, ash, lapilli (small stones), pumice and volcanic bombs. Also included is the incandescent cloud of gas, known as a *nuée ardente*, that sometimes accompanies eruptions. The capital of the island of Martinique, Saint-Pierre, was destroyed by a nuée ardente from the volcano Mt. Pelée in 1902. Nuée ardente forms a deposited material known as an ignimbrite.

pyrophytes are plants that have adaptations that enable them to withstand fire. This usually consists of bark that is fire-resistant. Examples of pyrophytes are the baobab tree and the acacia, both of which are typical of *savanna* regions. For some plants fire is required before they can regenerate. In Australia, for example, plants such as banksia need the fire for their woody fruit to open and thus regenerate.

Q

quadrat: a frame, usually a square, enclosing an area of known size, and used mainly for *sample* surveys of vegetation and surface deposits. A grid can be inserted by using wire or string, and this will provide sampling sites, particularly for a *systematic sample.*

quality of life is measured on an index devised in the 1980s and known as the Physical Quality of Life Index (PQLI). The PQLI is the average of three characteristics: *literacy, life expectancy* and *infant mortality.* Each feature is scaled from 0 to 100, with the lowest country in a category given 0 and the best 100. The three "scores" for each country are added and the average is found. This figure is the PQLI for that particular country. Almost all the countries that are normally considered to be in the *First World* have a PLQI of 90 or over.

quarrying: the removal of rocks for commercial purposes from large and open surface workings. Quarrying is part of the *primary sector* of economic activity. Although they can be important assets in some areas, quarry workings are not without their critics:

Pros:
- provides employment in many rural regions (where there may be few alternatives)
- brings revenue into rural regions
- could improve local *infrastructure*
- may be a factor in reducing *rural depopulation*
- provides important *raw materials* for other industries such as cement, steel and chemicals
- can reduce foreign spending on imported raw materials

Cons:
- visual *pollution* of quarry scars and waste tips, etc.
- noise from blasting and excavation
- very dusty in some places, both from quarrying activities and from vehicles
- increased vehicular activity on narrow rural roads.

quartile: when values are ranked in order to calculate the *interquartile range,* the first necessary step is to find the quartiles. Once the *median* has been found, the quartiles represent the median of the values above the median (upper quartile) and the median of the values below the median (lower quartile). For example, in a ranked list of 19 values, the upper quartile will be the 5th figure and the lower quartile the 15th.

Quaternary period: the latest period of geological time spanning the last 2 million years. It covers both the *Pleistocene* (which includes the most recent *Ice Age*) and the *Holocene* (postglacial period).

quaternary sector is a sector of economic activity that follows the tertiary (some believe it is part of the *tertiary sector*). This covers activities such as training and *research and development.* Industries involved include high technology (*hi-tech*) and information services.

questionnaire surveys involve both questions and answers and are used for obtaining opinions, views, ideas and information about the way people behave. Questionnaires can be completed in the street, door-to-door, by post or on the telephone. For geographical surveys, the best questionnaires involve a few short and uncomplicated questions that produce clear and precise information. Good questions ask people for the pattern of their behavior, not how they think they behave. For example, it is better to ask a shopper how many times he or she has been to the central business district that week, not "how often do you shop in the central business district?"

quota: the introduction of a set figure of production or the amount of import commodities that a country will allow in from certain sources.

R

radial drainage is represented by a pattern of streams radiating out from a central point. It results when land is uplifted to produce a dome structure with streams developing from the highest point. Radial drainage patterns can be seen at a variety of scales, for example, on a slag heap next to a coal mine, or on a volcanic cone.

radiation: the emission of electromagnetic waves from a body such as the sun or the Earth. An important feature of these waves is their wavelength, as this gives rise to different forms of radiation:

- shortwave radiation – emitted by the sun with its very high surface temperature as *insolation*
- longwave (terrestial) radiation – emitted by the surface of the Earth. This is caused by the Earth's surface absorbing insolation, and converting it into heat. The ground, warmed in this way, reradiates the energy back to space as longwave radiation. However, much of this radiation is absorbed by the atmosphere, thus warming it from below. (See *insolation, energy budget [the Earth]*.)

radiation fog results when a body of moist air, in contact with the ground surface, is cooled to its *dew point*. This commonly occurs at night under cloudless anticyclonic conditions with only a light breeze blowing. Because the sky is clear, the ground surface cools rapidly by *radiation*, and in turn cools the layer of air immediately above it. Once the dew point has been reached, *condensation* occurs. The cooled lower layer of air is stirred by the light wind so that it cools the air above it to its dew point, and the fog grows deeper. Radiation fogs commonly occur under *temperature inversions*, which prevent the air from rising. They may persist for several days in winter if the sun is too weak to disperse them. In industrial districts, an extreme form of radiation fog called a *smog* may occur.

raindrop impact: the erosional effect of raindrops falling in an intense storm. These droplets can have a very large kinetic energy, and on contact with a loose soil surface may cause a splash of particles both vertically and horizontally for some distance. On a flat surface, the particles merely change position, but on a slope the net transfer is downslope. This is because the particles splashed down the slope have longer trajectories than those splashed up the slope.

Raindrop impact can have two significant effects:

- larger particles of sand are left on the slope, whereas finer particles are moved down the slope
- finer particles may be removed from around a larger pebble or stone. This produces a column of finer material capped by a protective stone, called an earth pillar.

rainfall: droplets of water large enough to fall to the ground. Their diameter is usually between 0.5 and 5 mm, but drizzle is made up of smaller droplets. Rainfall is produced by *precipitation processes.*

rainforest: the name given to a vegetation type that can be found in the Amazon Basin of South America, the Zaire Basin and the Guinea coast of Africa, and parts of Southeast Asia, Indonesia and northern Australia. The climatic type of these areas is known as the *equatorial climate.*

The main characteristics of the rainforest vegetation are:

- tall evergreen trees (30–50 m) grow close together, forming a high canopy of leaves, called emergents. The trees have very few branches except at the top
- continuous growth occurs with the constant high temperatures and rainfall. As old leaves die and fall, new ones grow giving an evergreen appearance
- smaller trees and shrubs form additional leaf canopies at lower heights
- many different species of trees are found, with hardwoods such as rosewood, mahogany and ebony being common
- leaves are thick and leathery so as to be able to withstand the strong sunlight, and they often have drip tips to enable them to shed excess water
- the tall trees have buttress roots to give them support
- the dense leaf canopy shades the ground so that there is little undergrowth except near rivers and clearings where light can reach the ground
- lianas grow from the shaded ground and reach the light in the leaf canopies by climbing the larger trees
- *epiphytes* grow on the taller trees and occasionally have roots hanging in the air
- *saprophytes* are found in large numbers on the dark forest floor
- parasitic plants live on the trees of the forests and feed from them.

Rainforests form the habitat for a huge number of birds, insects and other animals. They are, however, very sensitive and fragile ecosystems. Their existence relies on the rapid recycling of nutrients, but once the cycle is broken by human interference, the forest will have difficulty reestablishing itself.

Original areas of tropical rainforest

raised beach: a coastal landform produced by the land rising relative to the sea. As the land rose, former beaches were raised above the influence of the waves, and they are frequently accompanied by a raised wave-cut platform, backed by relict cliffs. The existence of all of these features is evidence that the *isostatic readjustment* of land was not constant. Sea level had to be unchanging for a period of time to allow their creation. Seashells found on the raised beaches can be dated to indicate the age of the beach.

ranching: a system of agriculture based on the rearing of *livestock*, usually cattle or sheep, which earns the lowest net profit per hectare of any type of commercial farming. It is practiced in remote or difficult areas of the world on an extensive basis. Examples include sheep farming in central Australia, and Patagonia and cattle ranching in the Great Plains of the USA, the Argentinian Pampas and northern Australia. Large areas of land are needed to supply the animals with sufficient grass to feed on, and consequently it is found in areas of low population density. Each farm is a very large unit, often over 100 km^2, but little additional capital is used. The output per farm worker is high, with the demand for labor being seasonal, particularly for sheep farming. Ranching is dependent on good communication systems for exporting the produce, as well as the food preservation methods of refrigeration, freezing and canning. Many ranches are now owned by business interests and run by managers on the site.

R and D: see *research and development.*

random means due to or of chance; therefore no pattern should be detectable in any situation. In *sampling*, for example, a random sample is one that shows no bias and in which every member of the population has an equal chance of being interviewed or used. Random samples are usually obtained by using *random numbers.* In describing the distribution of settlement in an area using *nearest neighbor analysis*, a result that indicated a random distribution of settlement is one where there is no pattern to be seen in that distribution.

random numbers are numbers that have no pattern to them if used systematically. They could be drawn from a hat, but more probably are generated by computers and compiled in random number tables.

range: the term has two meanings:

- in statistics, this is a very simple method of determining the degree of *dispersion* of the values about the *mean.* It involves calculating the difference between the highest and lowest values, thus emphasizing only the extremes, but indicating nothing about the remainder of the values
- the maximum distance that people are willing to travel to obtain a good or service. This will depend on the type of good or service, its value, the availability and type of transportation, and the frequency of need for the good or service.

rank-size rule: an attempt to find a relationship between the population size of settlements in a country or region. The rule, promoted by George Kingsley *Zipf*, states that the size of settlements is inversely proportional to their rank. In other words, the second largest settlement should be half the size of the largest city, the third largest settlement, one-third the size, and so on. It is usual to plot the relationship on *logarithmic scales*, which, if the relationship is perfect, will give a straight line relationship on the graph.

Where the largest city is very much larger than the second, this is known as *urban primacy*, e.g. in Argentina. Some countries have two cities of almost equal size and this is known as a binary distribution, e.g. in Australia.

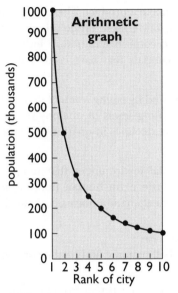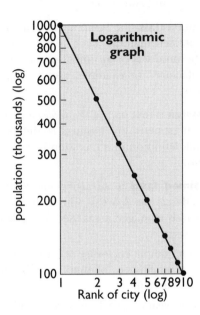

Rank-size rule

rapid transit systems are methods of moving people around large urban areas, and so avoiding the use of the car. Such systems involve light railways, underground systems and trams.

rationalization means reorganizing to increase efficiency. The term is mainly used when cutbacks in costs are needed in order to reduce an organization's break-even point. This may be achieved by:

- closing one of the company's factories and reallocating the production to the remaining sites
- closing an administrative department and delegating its tasks to the firm's operating divisions
- removing a layer of management.

The term is often used by companies as a euphemism for laying people off.

Ravenstein, Ernest George: he presented a paper in 1885 entitled "The Laws of *Migration.*" His main points were that:

- most migrants travel only short distances and with increasing distance, their numbers decrease (*distance decay*)
- most migrations produce a compensatory countermovement
- migration tends to occur in stages and with a wave-like motion
- urban dwellers are less likely to move than those in rural areas

- females tend to migrate more than males within their country of birth, but males are much more likely to be involved in international migration
- the major direction of migration is from rural to urban areas
- long-distance migration is more likely to be to large centers of industry and commerce.

raw materials: the *inputs* needed by a firm to make a product. These can be totally unprocessed materials such as iron ore, wheat, raw cotton, crude oil, etc., or they can be products of an initial process that have come from another industry. Car manufacture, for example, uses a range of materials that have come from other sources.

recession is that part of the *trade cycle* characterized by falling levels of demand, very little investment, low business confidence and rising levels of unemployment. The official definition of a recession is two successive declines in quarterly *gross domestic product.*

reclaimed land is an artificially created coastal environment that ranges from ditched *salt marsh* areas, where there is very little extra human interference, to intensively managed areas used for agriculture. Reclamation normally occurs in one of two ways:

- simple enclosure of former intertidal areas
- enclosure, followed by infill with sea bed material and finally pumping the area dry.

In recent years, coastal reclaimed land has come under threat from an increase in agricultural intensiveness, such as the use of *fertilizers* and *pesticides.* It is also under much greater pressure from increases in human activity such as camping and birdwatching.

recovery is that part of the *trade cycle* characterized by rising levels of demand, some investment, improving business confidence and falling levels of unemployment.

recreational activity includes any activity that is undertaken voluntarily for personal enjoyment in leisure time. This could include home-based hobbies and pastimes such as gardening, listening to music or watching TV, although generally the term relates to activities undertaken away from the home. Geographical research into leisure activity usually distinguishes between recreation, defined as pursuits that involve absence from home for less than one day, and tourism, which involves a longer absence.

recreational forest: an area of woodland, either privately owned or managed by a government body, that is open to the public for leisure activities. These could include nature/woodland trails, cycle routes, hides for birdwatching, and visitors' centers with on-site rangers. In some forests provision for visitors may extend to camping facilities or chalets/log cabins for rent. It provides an opportunity to manage the area for both economic and controlled leisure activity.

recyclable: any item that, after some form of processing, has the potential to be used again.

recycling is the reuse of materials that would normally be discarded after use. It has been encouraged by the threat of exhaustion of finite, *nonrenewable resources*. Scrap metals such as steel and aluminum have been recycled because reprocessing consumes less energy, and in the last 10 years there has been a great increase in the recycling of waste paper, glass and plastic. Pressure groups have made the public more aware of the problem and local governments have established recycling collections and centers.

redevelopment is the renewal and renovation of urban areas that have become run-down or derelict. This may be achieved by improving existing buildings through refurbishment or by widening roads to increase accessibility. Alternatively, buildings may be demolished and new structures built to replace them.

redlining is the demarcation of areas of a city in which financial organizations are unwilling to lend money for house purchase. These areas are perceived to be in decline and are regarded as areas in which lending money to potential purchasers would be a high risk. Banks, real estate agents and mortgage brokers effectively starve these areas of investment, discriminating against house owners in these areas and contributing to a general downward spiral in these districts. Redlining requires higher down payments on mortgages and imposes higher interest rates, effectively denying house purchase to low-income families. As housing deteriorates the neighborhood declines; home improvement loans are denied and property insurance becomes very expensive, even if it can be obtained. Businesses and services begin to fail and house prices decline rapidly. Redlining may help to inflate house prices in those areas where institutions are willing to invest. Although it is a legitimate practice, it does perpetuate decline and has important consequences for the social geography of the urban area.

reduction is a chemical process in which oxygen is removed, for example when ferric irons are changed to a ferrous state. Under waterlogged conditions, the pore spaces remain filled with water and the soil becomes deoxygenated. The iron in the soil becomes reduced and the soil has a greenish or bluish tinge, characteristic of *gleying*, which develops under *anaerobic* conditions. Reduction is the reverse process to *oxidation*.

refraction is the process by which waves change direction as they approach a headland, turning so that their crestline becomes more parallel to the shore. This concentrates wave energy and erosive power on the projecting headland where cliffs, arches and *stacks* may develop and reduces wave energy in the bays where deposition occurs to produce a bay head beach.

refugee: a person who, owing to fear of persecution, is outside the country of his or her nationality and is unable or unwilling to return to that country. This fear of persecution may be due to race, religion, nationality, or membership of a particular social or political group. This excludes people who may be forced to leave their homes but remain in their own country; these are termed "displaced persons." In recent years, terms such as economic, ecological and environmental refugees have appeared in the media, but as these people are simply seeking better economic opportunities they are not technically refugees and governments tend to regard them as ordinary immigrants who do not qualify for any special consideration with regard to political asylum. Escaping from poverty, crop failure or natural hazards is not the same as persecution; however, it is often difficult to distinguish the genuine refugee from the migrant who is merely aiming to improve standard of living. There are four main areas of refugee groups: Mexico and Central America, Northeast Central Africa (Uganda and Sudan), the Indian subcontinent and Iran, and Southeast Asia.

refurbishment: improvement to existing urban areas through investment in better road access or modifications to both the interior fittings and the external appearance of buildings as part of urban *redevelopment.*

regelation: the refreezing of water into ice on the downglacier side of an obstacle. When the underside of a *glacier* moves over an obstacle, *pressure melting* takes place on the upglacier side of that obstacle. This water acts as a lubricant enabling the glacier to move over the obstacle. However, on the downglacier side, the pressure is reduced, and the water reverts back to ice.

regeneration: the investment of capital and ideas into an area in order to revitalize and renew its economic, social and/or environmental condition. In recent years, the most common type of area in the USA to be regenerated has been the "*inner city.*" Key elements have been slum clearance and housing renewal, new industrial growth and development, improvements to transportation systems and environmental improvements. Private sector investment has been encouraged, instead of funding from local or central government sources.

region: an abstract concept of an area of a country, or across a group of countries, that has quite distinct boundaries, and within which there are generally similar characteristics. The notion of a "region" has been of interest to geographers for many years, but no simple definition has emerged. A region may be defined by its:

- physical characteristics – a similar relief (for example, the Paris Basin), or a similar climate (for example, the monsoon region)
- economic characteristics – for example, Silicon Valley in California
- functional interrelationships – for example, the city region, for example around Los Angeles, within which people travel, work, and obtain services.

regional contrasts arise from the inherent differences between varying parts of a country. No country has an *isotropic* surface – they all have variations in physical characteristics and in the distribution of resources. In addition, historical and political factors have resulted in the use and misuse of individual parts of a country. Some regions develop at a greater rate than others, and consequently other regions become exploited, neglected and more backward. Some economic theories have referred to regional contrasts by the concepts of the *core* and *periphery.* (See Gunnar *Myrdal* and *cumulative causation.*)

Regional contrasts may be measured by a wide range of economic and social indicators. These include unemployment rates, per capita incomes, rates of outmigration, education standards, and healthcare provision.

regional policy is the attempt by a government to redress imbalances of economic development and social and environmental conditions within a country by stimulating investment in the less prosperous areas.

Regional contrasts are regarded by some to be undesirable because:

- social unrest may arise in the less prosperous regions
- economically, national output is not maximized if there are both depressed "periphery" areas and overcrowded "core" areas
- politically, votes may be lost in the less prosperous regions.

Regional policies are not viewed as being successful by all. Some people argue that they have had little long-term effect – the regional contrasts tend to remain. Firms tend to migrate following the available funding and other incentives with the result that few "new" jobs are created. Substantial funding from central government is required, and this has led to an increased investment from the private sector in recent years. The need to subsidize failing industries and failing regions is no longer seen to be desirable.

regolith: the collective name for all of the material produced by *weathering* extending down from the ground surface to the unaltered bedrock.

regression: a more accurate method of showing the relationship between two variables than the *best-fit line* on a *scatter graph*. The procedure is as follows:

- calculate the *mean* values for both the dependent (y-axis) variables and the independent (x-axis) variables, and plot their location on the scatter graph as shown – point M
- draw a line parallel to the y axis through point M
- calculate the mean values for both the points to the left of this line, and for the points to the right of this line
- plot these two additional mean values on the scatter graph as shown – points ML and MR
- draw a line connecting the points M, ML and MR on the scatter graph. This regression line summarizes the characteristics of the scatter graph.

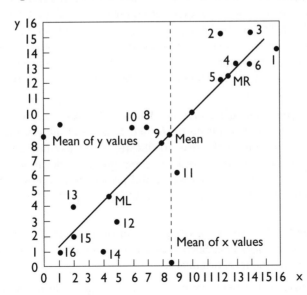

The construction of a regression line

Reilly, William J. produced a model of retail gravitation similar in nature to that of Newton's law of gravity. Reilly stated that two centers attract trade from intermediate places in direct proportion to the size of the centers and in inverse proportion to the distance between them. In other words, larger centers attract more trade than smaller centers, and the amount of trade with a smaller place increases the nearer that smaller place is to the larger center.

This model can also be used to define the sphere of influence between competing centers by using the breakpoint formula. If there are two towns A and B, and B is smaller than A, then the distance of the breakpoint from B is:

$$1 + \frac{\text{distance from A to B}}{\sqrt{(\text{population of A}/\text{population of B})}} =$$

Both of these aspects of Reilly's model assume that the larger the town the stronger its attraction, and that people always go to their nearest center. These assumptions may not be true as:

- there may be difficulties in reaching the larger town, and expensive forms of car parking
- the smaller center may be easier to reach, safer and less congested
- perceptions, positive or negative, of a place may override logical considerations.

rejuvenation: a fall in sea level relative to the level of the land, or a rise of the land relative to the level of the sea, enables a river to revive its erosional activity. The river will adjust to the new base level, at first in its lowest reaches, and then progressively inland. In doing so a number of landforms may be created: *knickpoints, river terraces* and *incised meanders.*

relative humidity: the amount of water vapor present in a body of air relative to the maximum amount of water vapor it could contain at that temperature and pressure. It is usually expressed as a percentage – a relative humidity of 100% means that the air is saturated. Since the amount of water vapor that a body of air can contain increases as its temperature rises, relative humidity is inversely proportionate to temperature. As the temperature of a body of air rises, its relative humidity decreases, and vice versa.

relict landscapes contain features or elements that are no longer being put to active use for their original purpose. Disused canals, old mine workings/quarries, abandoned factories, old dock areas, fortifications and mounds are all legacies of the past that can be studied in historical geography in order to gain some understanding of the connection between past and present human environments. Some of these relict features have become significant again as part of the growing heritage industry.

remembrement: the process of consolidation of agricultural holdings that have become excessively *fragmented.* Such a policy was promoted by the government of France, from where the term originates.

remote sensing is a general term covering a range of techniques that can be used to study features on or near the Earth's surface from a distance above the surface. Landforms, vegetation, land use and weather systems can be studied through aerial survey and the use of airborne electronic devices. Since the early 1970s satellites have been used to monitor changes on the surface and in the atmosphere. *Landsat* provides complete coverage of the Earth as the orbit changes slightly on each revolution. Some satellites such as Meteosat follow a fixed orbit and therefore remain above the same path; these are termed *geostationary satellites.*

rendzina: an intrazonal soil that develops where *limestone* or *chalk* is the *parent material* and under a grassland vegetation. The upper horizon of the soil is rich in *humus* and, given the nature of the parent material, the *pH* is between 7 and 8. Limestone and chalk when weathered leave little insoluble residue; therefore such soils tend to be thin.

renewable resources are *resources* that are either a flow of nature or living things, and can thus be used repeatedly. Living things such as forests are renewable as long as they are not used faster than they are replaced. In energy terms, renewable resources include *wind, tides, waves,* water power (*hydroelectric power*), *solar energy, geothermal, biomass,* and *ocean thermal energy.*

report: the conventional method of presenting precise information. A report may be used to convey an assessment of a situation or the results from qualitative and/or data analysis. It should have clearly stated aims and be tightly focused on the subject under investigation. To be successful, it should be easy to read.

research and development (R and D) means scientific research and technical development. This is directed at improving the product, rather than finding out what the customer wants and thinks (known as market research). With products having shorter lives before they are replaced by other types of products, R and D has taken on a greater significance, as companies have a need for further research in order to keep ahead. Companies also need to have some high-yielding products that will generate capital, without much present investment, in order to finance R and D. Such products are known as "cash cows."

residential segregation: the grouping of people with similar characteristics into separate residential areas. (See *social segregation.*)

residential type refers to the type of housing tenure in an urban area; traditionally this has taken one of three forms: owner-occupied (owned outright or purchased on a mortgage or other loan); privately rented (from a private landlord); and property rented from the government.

residual: a point on a *scatter graph* that lies some distance away from the best-fit trend line; this may be described as an anomalous value. A positive residual lies above the trend line, i.e. the *dependent variable* has a higher value than would be expected; a negative residual, below the line, indicates a lower value than would be expected.

resource: any feature of the environment that can be used to meet human needs. Traditional approaches consider that it is the act of exploitation that converts the commodity into a resource, i.e. they emphasize the use of the commodity. More radical approaches stress the exchange value of a resource. On the commodity markets, profits can be made without doing anything to a resource – owners buy at one price and sell on at another without perhaps ever seeing the resource or even taking ownership of it in the sense of storing it. Such speculative behavior, which also applies to purchase of land in anticipation of its future potential value, is viewing resources in a different way. Resources may be classed as *renewable*, flow resources, or *nonrenewable*, stock resources.

resource depletion: the consumption of *nonrenewable*, finite, resources, which will eventually lead to their exhaustion. This has led to greater awareness of the need for *conservation* and *resource management*.

resource management involves the control of resource exploitation and use in relation to economic and environmental costs. Sensible management of a resource includes some initial survey to determine the extent or quality of the resource, for example, soil fertility or land capability. This is followed by the development of strategies that might be used in the planning and exploitation stage, assessing possible conflicts that could arise if the resource use impinges on the interests of other groups. Resource management has to try and balance these conflicting interests and perceptions, considering both economic aspects and conservation issues.

retailing: the sale of goods and services to the public. The term usually includes high- and low-order shopper goods, such as food, clothing and consumer durables (fridges, washing machines, etc.) and is often extended to include consumer, professional and financial services such as hairdressers and banking. Fast food outlets, restaurants and entertainment may also be included.

ria: an irregular coastal inlet formed as a result of a relative rise in sea level causing submergence of a former valley system. A ria is deepest at its mouth and becomes progressively more shallow inland; in cross-section it resembles the former valley profile with some alluviation along the bed of the original channel due to the reduced energy along the river following the sea level rise. The heads of the tributary inlets may also be undergoing infilling by alluviation and the development of tidal mud flats.

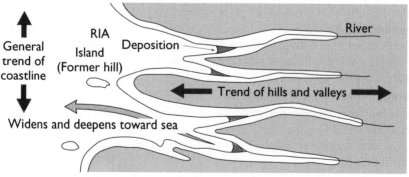

Ria coastline

ribbon lake: an elongated lake occupying the floor of a glaciated valley or trough. They occupy rock basins that have been produced by selective overdeepening; postglacial melting has resulted in a lake that may also be dammed by depositional material.

Richter scale: a system devised in 1935 to measure the magnitude, or total energy release, of an *earthquake*. It is a *logarithmic scale* that extends up to 10; an earthquake at Richter Scale 5 has a magnitude ten times greater than one at Scale 4. Ground tremors record a magnitude of 2 and damage to structures such as buildings occurs at Scale 6 and over. The largest shock recorded so far had a magnitude of 8.9.

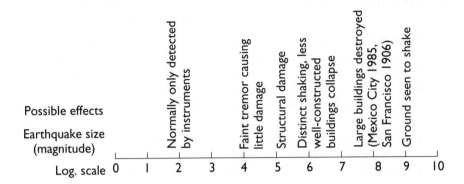

Richter scale

riffle: see *pools and riffles*.

rift valley: a structural landform resulting from the downthrow of rock strata between parallel faults. The best-known example is the East African Rift Valley, which extends from Mozambique in the south through East Africa and northward into Jordan, a distance of 5500 km. In some areas the inward-facing fault scarps are 600 m above the floor of the rift valley, and there are step faults on the valley side that indicate that a series of parallel fault lines have moved to cause a lowering of the floor section. The relief is further complicated by the escape of volcanic lava within the rift area; the valley is thought to be the result of tension in the underlying crustal rocks and it may indicate that the continental crust of the African Plate is separating to produce a divergent or *constructive boundary*.

rilling develops when a series of roughly parallel small channels form on a slope. They are closely spaced and their maximum size is usually up to 2 meters wide and 50 cm deep. They only discharge water during and immediately after heavy rainfall, but the rapid run-off can cause erosion and rills may produce *gullying*.

Rio Earth Summit: met in June 1992 to find ways of coping with the permanent environmental damage that has accompanied recent economic developments. Specifically, this meant devising strategies for coping with the effects of *greenhouse* gases, the accelerating loss of biodiversity and concerns over the environmental consequences of rapid population and industrial growth in the *Third World*. The Summit produced Agenda 21, a blueprint aimed at cleaning up the global environment and encouraging environmentally sound developments. Particular areas that were targeted included:

- poverty alleviation and environmental health in the Third World
- family planning and expanding job and educational opportunities for women

- reduction of *soil erosion* and the promotion of environmentally sensitive agricultural programs
- protection of natural habitats and biodiversity
- research and development on noncarbon energy alternatives in order to reduce greenhouse gases and to avoid *climatic change*.

risk assessment: a way of judging the degree of damage that an area may experience as a result of a *volcanic eruption, earthquake* or any other possible *hazard*.

river capture is where the headwaters of one river system capture those of another river system. A river with a steeper gradient, flowing over less resistant rock strata or with greater rainfall within the basin may extend itself by headward erosion to capture part of another stream. This is sometimes referred to as stream piracy. The main features that result from this action and indicate that it could have taken place are:

- elbow of capture – the point at which the headwaters of the river that has been captured sharply change direction
- wind gap – a valley where the captured stream used to flow
- misfit stream/river – the river that has been captured has a much reduced flow in the old lower course so that it now occupies a valley that is much larger than the present stream could have cut.

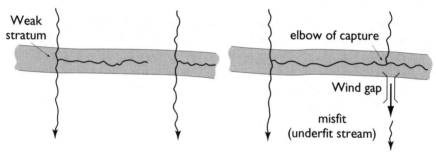

An example of river capture

river cliff: the steep slope that characterizes the outer bend of a river *meander*.

river deposition: when a river loses the ability to transport its *load* (*competence*) deposition occurs. This includes the following circumstances:

- when the river slows down, e.g. entering the sea, at the inside of a *meander*, when it overflows its banks or there is a sudden change in gradient
- when there is a reduction in volume during times of drought
- when the load is suddenly increased, e.g. *landslide*, entry of large tributary.

Rivers generally deposit the heaviest part of their load first, grading over time and space to the finest.

river erosion: a collection of processes that contribute to the wearing away of the banks and bed of the river and thus contribute to the *load* for transportation (see *Hjulstrom curve*). The main processes are:

- *corrasion* – the use of the load to wear away the banks and bed
- *attrition* – the wearing down of the load

- *hydraulic action* – the sheer force of the water itself. Also included here is *cavitation*, where the collapse of bubbles of air within the water sends out shock waves that weaken the banks
- *solution* – related to the chemical composition of the water and the geological make-up of the river valley.

river regime is the term used to describe the annual variation in *discharge*. Rivers in differing climates experience very different regimes.

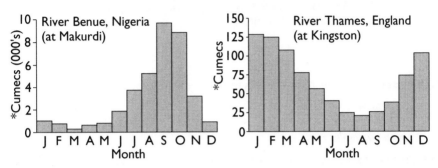

Examples of river regimes

river terrace: a remnant of a former *flood plain* that, after *rejuvenation* of the river, has been left at a higher level. The River Thames has created terraces in its lower course that are now occupied by parts of London.

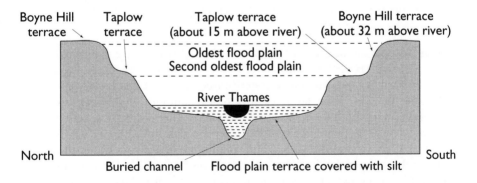

The terraces of the River Thames, London

robotics is the science of using robots in production processes to replace people, especially where such processes are monotonous or hazardous.

roche moutonnée: a protruding knob of bedrock found along the floor of a glaciated valley. One side of the rock is gently sloped and smooth (the stoss side), whereas the other (the lee side) is steep and uneven. It was formed by the action of a *glacier*, which, on encountering a more resistant protusion of rock beneath it, scratched and polished the stoss side as it slid over it, and then plucked out rocks loosened by *freeze-thaw* action and *pressure release* from the lee side. (See *plucking* and *regelation.*)

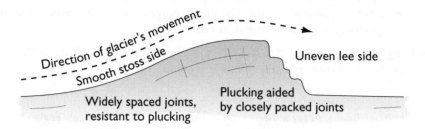

Roche moutonnée

rock type: the classification of rocks into three main groups:

- *igneous rocks*
- *sedimentary rocks*
- *metamorphic rocks.*

roll on–roll off: a method of sea transportation that allows trucks loaded with freight and other vehicles to drive onto a ship, and off again at the destination, thus eliminating the need for cranes. The time taken to load and unload a ship is reduced, and so it can spend more time at sea, increasing efficiency. Consequences of this have been:

- the great reduction in the need for a dock labor force
- the rise in importance of ferry ports
- the development of fleets of specially designed sea-going ferries.

ro-ro: see *roll on–roll off.*

Rossby waves: the pattern of flow of the winds (known as the Upper Westerlies) in the higher parts of the atmosphere. It is known that in the upper atmosphere winds blow around the planet in a westerly direction, but follow a wave pattern. The waves stretch from polar latitudes to tropical latitudes, and there are usually between four and six of them in each hemisphere. The reason for their existence is not clear, but some people believe that they are due to the upper air flow being forced to divert around the great north-south mountain ranges of the Rockies and Andes in the northern and southern hemispheres respectively. Once a wave motion has begun, it is perpetuated around the planet. The waves have considerable variation in amplitude during a year. It is this variation of amplitude, together with their relatively static locations, that have a significant effect upon the creation of both low pressure and high pressure areas on the surface.

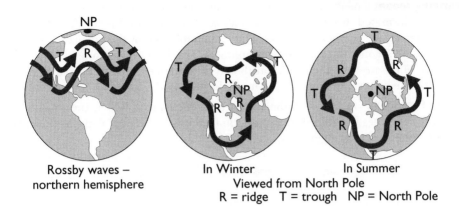

Rossby waves –
northern hemisphere

In Winter
In Summer
Viewed from North Pole
R = ridge   T = trough   NP = North Pole

Rossby waves

Rostow, W. W.: put forward a model of economic growth for a country that identifies five stages of development:

1 traditional society – a subsistence economy based mainly on farming with limited technology or capital to process raw materials or to develop industries
2 preconditions for take-off – an injection of investment allows both agriculture to be more commercialized and a single industry to begin to dominate the economy. Technological developments cause a growth in infrastructure, including a transportation system
3 take-off – manufacturing industries grow rapidly, with growth being concentrated in one or two parts of the country. Improvements in the transportation infrastructure continue, together with some progress in social conditions. Employment in agriculture declines
4 the drive to maturity – a period of self-sustaining growth. Economic growth spreads to all parts of the country; more industries are developed, and urbanization takes place on a rapid scale
5 the age of mass-consumption – a period of rapid expansion of service industries, with a consequent decline in manufacturing.

The key element of this model is the need for an injection of capital, usually from an external source, for example an *economically more developed country (EMDC)*. This has taken place in many African and Asian countries, with little effect other than to cause the build-up of huge national debts. The model is now regarded as being too simplistic, as well as too heavily based on the experiences of Western countries that grew economically in the earlier part of the 20th century.

rotational movement: a slippage or slide along a curved plane. This type of movement takes place in:

- the *slumping* of debris down a slope, producing a tear or scar at the back of the slope, and a lobe of material at the foot of the slope
- the formation of a *cirque (corrie)* where the glacier is thought to pivot as it moves out of the hollow.

running mean: a method by which the overall trend of a set of values in a data set can be determined. It involves the calculation of the mean value for, say, the first three consecutive numbers in the set. This is then followed by the calculation of the mean of the second, third and fourth numbers. This is in turn followed by the calculation of the mean of the third, fourth and fifth numbers, and so on. This example is a three-item running mean, but it may be of any other number.

One of the main outcomes of the use of a running mean is that the impact of exceptional values is reduced. However, another is that the final items in the data set will not have a running mean of their own.

1974	1975	1976	1977	1978	1979	1980	1981	1982	Moving total	Running Mean	Middle year
6.1	2.0	2.6	3.3	1.5	5.7	29.2	3.6	0.4			
6.1	2.0	2.6							10.7	3.6	1975
	2.0	2.6	3.3						7.9	2.6	1976
		2.6	3.3	1.5					7.4	2.5	1977
			3.3	1.5	5.7				10.5	3.5	1978
				1.5	5.7	29.2			36.4	12.1	1979
					5.7	29.2	3.6		38.5	12.8	1980
						29.2	3.6	0.4	33.2	11.1	1981

3-year running mean

run-off is all of the water that enters a river and flows out of a *drainage basin*. *Throughflow*, groundwater flow and *overland flow* all contribute to run-off. It can be quantified by measuring the *discharge* of that river.

rural depopulation: the movement of people out from rural areas and into urban areas. This movement can be attributed to rural *push* and urban *pull factors*. They include:

- rural push factors – small fragmented plots of land that are too small to support a family; crop failures; the use of mechanization reducing the need for labor on the land; the clearance of the land by large landowners who wish to redevelop it for commercial purposes
- urban pull factors – the perception of a better lifestyle; the prospect of a job and cash wages; the higher quality of health and education services.

rural settlement includes isolated farmhouses and small towns. However, some of these are difficult to distinguish from urban settlements as townspeople increasingly move out into the countryside.

S

Sahel: the region of Central West Africa lying between approximately 10°N and 20°N. It includes the countries of Senegal, Mauritania, Mali, Burkina Faso, Niger and Chad. The Sahel is mostly desert margin and has suffered from a number of prolonged droughts in recent years.

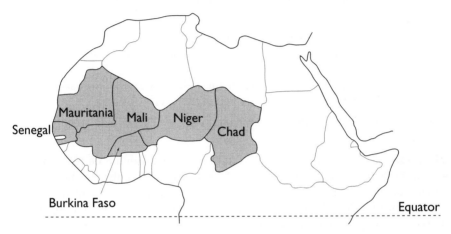

The Sahel countries of Africa

salinization occurs when *potential evapotranspiration* is greater than *precipitation* and when the *water table* is near to the ground surface. It is therefore a feature of areas with an arid or semiarid climate. As moisture is evaporated from the surface, water containing salts is drawn upward by *capillary action*. Further evaporation causes the deposition of the salts on the ground surface.

Salinization has become a major problem in some irrigated areas. Where the irrigation water is unable to drain away, waterlogging has taken place in the soil. The water table has risen up through the soil, bringing salts with it. The roots of plants that cannot tolerate saline conditions become affected and the plants die. The solution to this problem is to provide adequate field drainage together with a constant flow of irrigation water to flush the salts out of the soil.

SALR: see *saturated adiabatic lapse rate.*

Saltaire is a model industrial village and textile mill north of Bradford, England, built by the industrialist Sir Titus Salt. The village is a perfectly preserved 25-acre site of one man's utopia. Having built his fortune on the use of alpaca and mohair, Salt found that by the late 1840s his factory was too small to meet the demands of his new textiles. However, in 1849 a major cholera epidemic struck Bradford. Being a strict congregational Christian, he stated that "cholera was God's voice to the people," and so decided to build a better community for his workers. Saltaire was built between 1852 and 1872 and was modeled on the buildings of the Italian

Renaissance period – a period (in Salt's opinion) when both cultural and social advancement took place. Salt's mill emulated an Italian Palazzo and was the largest factory in the world when it opened. It was surrounded by a school, a hospital, a railroad station, parks, baths, washhouses, 45 houses for the poor, and 850 houses. The style and size of each house reflected the place of the head of the family in the factory hierarchy. Twenty-two streets were created, and all but two (Victoria and Albert Streets) were named after members of Salt's family. The church was the first public building to be completed, and there was not a single bar.

saltation: the process by which particles are lifted bodily upward and forward before returning back to the surface from which they started. This occurs in:

- desert areas where the wind picks up fine sands and moves them in the direction of the wind. The individual grains bounce along leapfrogging over each other, but rarely reach heights of more than 2 m above the ground
- rivers where the sheer force of the moving water bounces sands, gravels and larger particles along the bed of the river.

In both cases, impact with the returning surface may initiate further particle movements.

saltmarsh: an area of vegetated tidal mudflats located within an estuary or on the landward side of a *sandspit*. Fine sediment accumulates in the sheltered water, which is then colonized by a sequence of salt-tolerant plants as the level of the marsh is raised by further deposition. Most saltmarshes tend to have a well-developed series of creeks through which the tide rises and ebbs. The sequence of vegetation colonization that takes place on a saltmarsh is an example of the plant *succession* known as a *halosere*.

sample size: the number of respondents or objects in a research survey. The sample size is important as it needs to be large enough to make the data statistically valid. A sample of 20 people, for example, is so small that a different 20 could easily have quite separate views or behave in a completely different manner. When deciding on an appropriate sample size, the following should be borne in mind:

- the higher the sample size, the more expensive the research in terms of cost and time
- the lower the sample size, the greater the chance that *random* factors will make the result inaccurate.

sampling: to make statistically valid inferences, when it is impossible to measure the whole *population*, by selecting a group that will be representative of that population. Such a group is known as the sample. There are three main methods of sampling: *random, systematic* and *stratified.*

sandspit: an embankment of sand and shingle (pebbles) attached to the land at one end. The unattached end is often curved inward toward the land, and may have several recurved ridges known as laterals. The highest parts of a sandspit often consist of sand dunes. Here sand has blown from the beach, and has been stabilized by vegetation such as *marram grass.* Along the seaward side, ridges of shingle are frequently found, thrown up by the sea during storms. On the landward side, a *saltmarsh* may be found where fine silts and mud have been deposited by the mixing of fresh and salt water (*flocculation*).

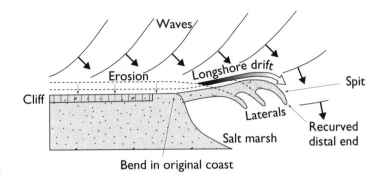

Sandspit

sandur (plural: sandar): the Icelandic term for the *outwash* plain found in front of the *ice margin*. A sandur consists of *fluvioglacial sands* and gravels crossed by meltwater streams.

saprophyte: a group of organisms, including some plants, that obtain *nutrients* from dead or dying organisms. Most fungi are saprophytes.

satellite photographs: images that can be transmitted from satellites, particularly useful in the fields of *weather forecasting* and land use analysis. This is one example of *remote sensing*. Photographs can be produced using ordinary light but most satellites have sensors that can detect different bands of energy emitted from the Earth. The *infrared* part of the electromagnetic spectrum is one of the measurements that can be taken and transmitted as a photograph.

satisficing: a term used to describe the acceptance of what is satisfactory instead of pursuing the best or maximum result.

saturated adiabatic lapse rate (SALR) refers to the rate at which the temperature of a saturated body of air falls as the air is forced to rise. The temperature fall is caused by heat loss due to the expansion of the body of air. It is lower than the *dry adiabatic lapse rate (DALR)* because as *condensation* takes place, *latent heat* is released, which compensates for some of the heat loss. However, the SALR varies because the warmer the air the more moisture it can contain and so the greater the amount of latent heat released following condensation. It may be as low as 4°C per 1000 m or as high as 9°C per 1000 m.

savanna is the name given to a climatic and/or vegetation type found in parts of *tropical* sub-Saharan Africa, the Brazilian Plateau and northern Australia.

The climatic type is also known as the Tropical Wet and Dry, or Tropical Summer Rain. The main features of the climate are:

- a hot and wet season with temperatures over 26°C, and heavy convectional rainstorms
- a slightly cooler dry season, with temperatures averaging 21°C, and little or no rain
- the length of the dry season decreases polewards, until the desert type climate prevails.

The dry season corresponds to offshore *trade winds* blowing across the area from a dry interior. The wet season occurs when the *intertropical convergence zone* moves polewards following the overhead sun. This causes low pressure to develop, and produces strong convection and intense rainfall.

The vegetation type is dominated by grassland with both evergreen and deciduous trees. The grass decreases in height polewards with an increasing dry season, and the cover becomes more sparse near to the desert margins with bare ground occurring in between tufts of spiky grass. The trees vary from being evergreen near to the equatorial margin, to being deciduous further away. They have adapted to the drought conditions (*xerophytic*) by having thick barks, small thorny leaves, and extensive root systems. The acacia also has an umbrella-like shape to shade the root area beneath. The baobab stores water within its large trunk.

Some people believe that this vegetation type exists because of the actions of humans who have fired the area consistently over thousands of years. The grasses can recover during the wet season, and the trees, such as the acacia and baobab, are fire-resistant (*pyrophytes*).

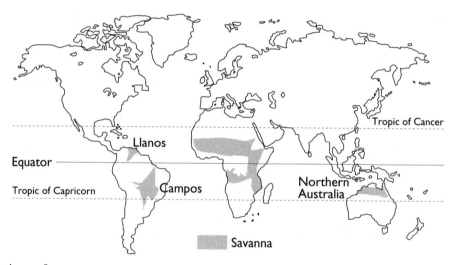

Areas of savanna

scale: the degree to which a map is smaller than the area that it represents. If the scale of a map is known, a measurement made on the map can be converted to its equivalent on the ground. The scale may be expressed as:

- a representative fraction – for example, the Landranger Series of the British Ordnance Survey has a representative fraction of 1 : 50,000. One centimeter on the map represents 50,000 cm on the ground, or 0.5 km
- a linear scale – which on the Landranger Series mentioned above would be a 2 cm length on the map for every kilometer on the ground. A linear scale is also usually subdivided into smaller fractions, such as tenths.

Scale may also be applied to the size of the area of study. It may vary from a local scale, to a national, regional, international or global scale.

scarp: the steep slope of a *cuesta* (*escarpment*). A *spring* is a common feature at the base of a limestone or chalk scarp.

scatter graph: a graph showing the relationship between two sets of variables by the distribution of dots. It is usual that the *dependent variable* is placed on the y-axis, and the *independent variable* on the x-axis. Dots are plotted onto the graph using the two sets of data as coordinates. The arrangement of the dots can then be examined to see if there is:

- a positive relationship – as one variable increases, so does the other
- a negative relationship – as one variable increases the other decreases
- no relationship – there is no recognizable pattern to the distribution of dots.

A positive or a negative relationship can be indicated by inserting a *best-fit line*. (See also *regression*.)

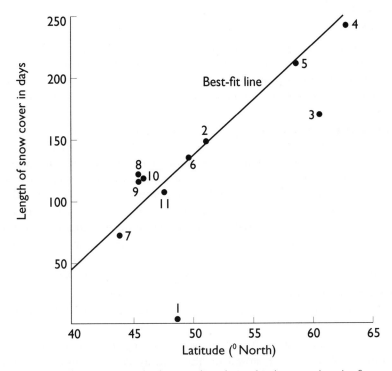

An example of a scatter graph: showing the relationship between length of snow cover and latitude for 11 towns in Canada.

Schengen agreement: a decision by a number of *EU* countries to remove all border controls such as passport and customs checks on their internal or shared frontiers by 1994. This followed the Schengen Accord, established in 1985, and was actually implemented in March 1995 when seven countries – France, Germany, Spain, Portugal and the Benelux countries – abolished border controls. Flights between airports in these countries are classed as "domestic," enabling business executives and tourists to travel without passports, immigration checks or other border formalities. The countries reluctant to join are concerned about the effect on movement of drug smugglers, international criminals, terrorists and illegal immigrants, although the participating states accept the need for much tighter "external" controls. The United Kingdom, in particular, was concerned that the agreement would cancel out the advantages of being an island in controlling illegal entry. Italy and Greece have also signed and expect to join and Denmark, Sweden and Finland have expressed interest. Ireland would like to join but has been deterred by the UK's position.

science park: an industrial complex located near to a university or research center with the intention of encouraging and developing *hi-tech industries.* Businessmen and academics can meet to identify and discuss practical applications of new technological research developments.

sclerophyllous refers to plants with drought adaptations to reduce the loss of water through *transpiration.* Specifically it applies to leaves that are thick, leathery and have thick cuticles. Such plants are typical of arid, semiarid and scrub environments such as those of *Mediterranean climatic* areas: *maquis, garrigue* and *chaparral.*

scree: angular rock fragments that accumulate on a slope below a steep rock face (free face) or summit. The debris, the result of physical weathering processes, may be of sufficient extent to produce a scree slope that covers the surface rock. They are common in glaciated mountain areas where *freeze-thaw* acts on steep bare rock outcrops.

sea floor spreading is the movement of oceanic crustal plates away from divergent or *constructive plate boundaries.* In the 1960s the mechanism behind the theory of *continental drift* was confirmed by examining the floor of the Atlantic Ocean, which revealed a ridge built up by successive eruptions of basaltic volcanic lava, similar in composition to the oceanic crust. Evidence showed that the sea floor becomes older with distance from the Mid-Atlantic Ridge, the point of extrusion along the divergent boundary. As new crust is created by the upwelling magma the ocean crust moves away from the boundary. The rocks also reveal evidence of magnetic striping and *paleomagnetism.* In the Atlantic the rate of spreading is about 1 cm per year. In the last 20 million years the Atlantic has increased in width by over 400 km.

sea level change occurs when the level of the sea in relation to the land varies. Either the land or the sea may change its level. Sea level changes can be classified as being either *eustatic* or *isostatic.*

Eustatic sea level changes apply to the whole world. During times of major glaciation, large volumes of water were stored on the land as ice. As a result sea level worldwide fell by 100–150 m. However, since the great ice sheets have begun to melt, sea level on a world scale has risen as the water has returned to the sea.

Isostatic sea level changes concern only certain parts of the world. During the major glaciations, the weight of the ice depressed the Earth's crust immediately beneath the ice. This caused a local relative rise in sea level.

Since the ice sheets have melted, the Earth's crust, which had been beneath the ice sheets, has begun to rise. This has caused a local relative fall in sea level.

These two types of change often appear to work against each other, although it would appear that isostatic changes are slower to occur then eustatic. Consequently, parts of northwest Scotland and the Gulf of Bothnia in Scandinavia are still experiencing slight falls in sea level some 10,000 years after the last ice sheets disappeared.

secchi disc is a black and white colored disc used in studies of water *pollution*, which is lowered into water until it is no longer visible (this depth is recorded) and raised again until it reappears (this depth is recorded). The average of the two is the secchi depth, which varies with the *algal* build-up and reflects the *trophic* status of the water.

secondary data: information that is derived from published documentary sources and has been processed, such as processed *census* data, research papers, textbooks, etc.

secondary forest: the vegetation that develops following the clearance of the original vegetation and recolonization through a *secondary succession*. The vegetation may not have reattained the climax stage and secondary forest represents a *seral stage* in the gradual succession back to the *climax community*. This is found within tropical rainforest areas where *shifting cultivators* have cleared the original vegetation.

secondary sector: the sector of the economy that is concerned with the processing of any primary raw materials and the resultant manufacture of products.

secondary succession: a sequence of plant succession on land that has previously been vegetated. It is characteristic of areas that have been cultivated/grazed and later abandoned, or where forest areas have been cleared by felling or as a result of devastation by fire. Because these subseres develop on sites that have already been biologically modified, in that they have some soil and organic matter, the succession is more rapid than in a *prisere*. The pioneer stage may be very short or absent, and the succession tends to be "telescoped" in time, being particularly rapid on formerly cultivated land where there may be improved soil conditions through drainage, plowing or the use of organic/inorganic fertilizers.

Second World: the communist (socialist) countries of Eastern Europe and the former *USSR*. The term was introduced after World War II as the Cold War between the capitalist *First World* and the Eastern Bloc developed. It has been of less significance since 1991 with the subdivision of parts of the USSR and its satellite states.

sectoral change: the variation in the relative proportion of the working population in each sector of the economy. As a country begins to develop economically there is a change in the balance of employment in each sector. Initially the *primary sector* dominates with a subsistent economy, but as processing of raw materials develops there is an increase in the *secondary sector*, and this continues as industry becomes more varied through *cumulative causation* and *linkage*. As the secondary sector increases the primary sector falls. The tertiary sector also starts to rise as opportunities emerge for financial and consumer services. As secondary industry becomes more efficient in its use of mechanization or robotization, workers are released and the tertiary sector dominates as both primary and secondary sectors decline. In many economically advanced countries, *deindustrialization* is leading to further decline in secondary employment and a marked change toward the tertiary and *quaternary sectors*.

sector model: see Homer *Hoyt*.

sediment: any material, ranging in size from fine clays to boulders, that has been transported and deposited by an agent of *erosion* such as water, wind and ice.

sedimentary rock: a rock made up of sediments deposited in distinct layers separated by bedding planes. They consist of material that is the debris produced by the breakdown of other rocks, *igneous, metamorphic* and older sedimentaries. These sediments accumulate in environments such as deep oceans, on the continental shelf, and in inland lakes and large river valleys. The overlying weight of sediment causes compaction, cementation and hardening, a process referred to as lithification. Rocks can be subdivided according to their dominant constituents:

- argillaceous – comprise fine material such as clays
- arenaceous – consist of cemented sands
- rudaceous – comprise coarse material, e.g. conglomerates
- calcareous – consist of calcium carbonate derived from the remains of shells and corals as in *chalk* and *limestone*.

sedimentation: the deposition of *sediment.* The term is usually applied to the accumulation of silts and sands in a water course where the deposition may raise the bed of the channel and pose a flood risk or where sediments infill reservoirs.

sediment yield: the total amount of *sediment* being transported by a river over a certain period of time. It is used as a measure of the rate of erosion operating in a drainage basin. It is calculated by measuring the sediment *load* (for practical purposes this is normally restricted to the *suspended load*) and dividing this by the area of the drainage basin. This is usually expressed as volume per unit area per year, i.e. $m^3 \, km^{-2} \, year^{-1}$ or tons $km^{-2} \, year^{-1}$.

It expresses the rate at which material is being removed, or eroded, from the basin. Studies of erosion rates in different environments produce conflicting results because it is difficult to allow for the effects of human disturbances, such as cultivation and deforestation, that assist erosion. Generally in the humid tropics, in undisturbed areas, the dense vegetation helps to maintain low rates between 15 and 20 $m^3/km^2/yr$, while in disturbed environments very high rates may be found. In the desert regions, rates may be as low as 1–2 $m^3/km^2/yr$. In the temperate and semiarid regions the rates are particularly high, ranging from 15–100 $m^3/km^2/yr$.

segregation is when certain groups of people live apart either because they are forced to do so or for economic and social reasons. Segregation of people can be based on race, wealth or age. (See also *social segregation.*)

seif: a type of desert sand dune aligned with the *prevailing wind* direction. It is usually asymmetrical in cross-section, and secondary winds, i.e. less frequent and less powerful winds, may help to shape this elongated dune. Some researchers suggest that seif dunes may be the result of migration and coalescence of crescent-shaped *barchan* dunes that become distorted by changing wind directions.

seismic: a term meaning "of an earthquake," as in seismic waves, seismic focus and seismology.

self-employment is when workers operate as their own bosses, either working freelance or with the permanent task of running their own businesses. Self-employment has various tax advantages over regular employment, especially in claiming expenses that can be offset against income tax bills. However, the self-employed get no paid vacation time or sick leave and no healthcare provision.

self-sufficiency is where a region/country does not depend on outside help in the supply of a particular commodity. A country, for example, that produced enough grain on its own farms to feed its own people, did not import any, and perhaps even exported some grain, would be said to be self-sufficient in that commodity.

separatism: the attempt by groups within a country to achieve greater autonomy, and ideally total independence, from a central government from which they feel alienated. These areas often have distinct languages and culture and are peripheral geographically to the *core* within the nation state. Separatism is common throughout the world, such movements varying from underground, violent and illegal organizations (ETA organization within the Basque provinces of France and Spain), to peaceful political parties (Scottish National Party/Plaid Cymru, Wales). Examples of such movements include:

- United Kingdom – Scottish and Welsh nationalism
- Spain – the Basque provinces and Catalonian nationalism
- France – Breton nationalism
- Canada – Quebec separatist movement
- Russia – Chechen independence (which led to the Russian invasion of 1994/95).

sere is a particular kind of plant *succession*. There are several major types recognized:

- *lithosere* – developed on bare rock
- *psammosere* – developed on a sandy surface
- *halosere* – salt water environment
- *hydrosere* – fresh water environment.

Within the succession each recognizable stage is known as a seral stage.

service sector: see *tertiary sector*.

sesquioxide: chemical compounds, which are common in many soils, resulting from the *weathering* process. They consist of the oxides of the two primary minerals, iron and aluminum.

set-aside: a scheme announced by the *EU* in 1988, by which farmers who agreed to take *arable* land out of production would receive compensation. The main features of the scheme were:

- farmers had to take at least 20% of their arable land out of production
- the land had to be left fallow (following strict guidelines on cutting and maintenance), turned into woodland or put to a nonagricultural use
- the agreements originally had to be over a five-year period.

In 1992 the minimum area was reduced to 15%, but all farmers in receipt of EU funding would be required to join the scheme. This highly controversial measure was seen by ministers as the only fast way to reduce the overproduction of cereals within the EU.

In 1995 the EU agreed an extension to the scheme that allows farmers to claim set-aside *subsidies* on woodland, heaths and other areas devoted to wildlife and *conservation* projects. Many farmers have been reluctant to take part in new conservation schemes, which, they claim, would make their farms unprofitable by making too much land unproductive.

Other general objections to the scheme have been:

- the rates of compensation are worth more to farmers in marginal rather than the core arable areas
- most farmers put only their poorest land into set-aside (which meant that they would be required to rotate the set-aside land)
- grain production would fall too much, meaning that reserve stocks could be too low for emergencies
- the scheme involved too much bureaucracy and the employment of more European civil servants!

settlement function refers to the main activities in a place. Function relates to the economic and social development of a center. Major functions include residential, commercial, agricultural, industrial, mining, religious, administrative, tourism/recreational, defensive, market center, transportation center.

settlement morphology refers to the pattern or shape of settlements. Morphology can also be linked with structure where detailed attention is given to the building type, age, construction and function within *land use* zones in the settlement. The pattern or shape made by the functional zones within a settlement is known as the functional morphology.

settlement site: see *site.*

settlement situation: see *situation.*

shanty town: see *squatter settlements.*

shape index: a statistical measure for analyzing the shape of any geographical area such as a county, a built-up area or a regional unit. The formula used is:

$$\text{shape index (S)} = \frac{1.27 \times A}{L^2}$$

where A is the area of the unit in square kilometers, and L is the long axis of the area drawn as a straight line connecting the two most distance points on the perimeter. The multiplier 1.27 is used so that a circle would produce an index of 1.0 with values ranging down to zero.

Maximum compaction occurs when $S = 1$, i.e. the shape is circular. This is mathematically the most efficient way of enclosing a given area. As an area becomes more elongated the index falls. The single figure, numerical index allows direct comparison of areas of different size and shape.

sharecropping: an agricultural system in which a farmer has to give part of his crop to the landowner rather than a fixed rent. It is not unusual for this to be at least 50 to 60%, which gives the farmer little incentive for improvement. It was a widespread practice in the cotton belt following the Civil War and the abolition of slavery.

shield area: see *craton.*

shifting cultivation is the system of farming that developed in areas of tropical *rainforest* where plots are created from the forest, cultivated and then abandoned. Plots would not then be reused for many years. It is also known as *slash and burn* and by various local names such as milpa in Latin America, ladang in Southeast Asia and chitimene in Central Africa (on the wooded savanna rather than rainforest). As a farming system, shifting cultivation is very energy-efficient and operates in close harmony with the environment. As the tribes do not work the soil for long periods, *humus* will build up sufficiently for reuse, making this a very *sustainable* form of forest management.

Much of this way of life has been destroyed with the destruction of the rainforest. In the Amazon Basin, for example, land is being cleared for cattle ranches, timber, *hydroelectric* schemes, highways, reservoirs and other commercial exploitation, particularly the attempts to introduce sedentary cultivators. As a consequence the shifting cultivators are forced either deeper into the untouched forest areas or onto designated reservations.

shingle is the name given to material of gravel or pebble size that accumulates and forms beaches and off-shore *bars*.

shopping center: see *mall* and *strip mall.*

shopping mall: see *mall.*

shrinkage of space: the result of the changes in transportation and communications technologies that reduce the *frictional effect of distance* on movement and therefore permit space to shrink. Some of the technological changes that have contributed to this are:

- in ocean transport, the development of giant oil tankers, *containerization* and the introduction of *roll on-roll off* methods
- in communications, satellites and their ability to transmit live television images and transfer business information.

sial: the upper, continental part of the *crust* of the Earth, which is composed predominantly of silica and aluminum (from which the term is derived).

significance testing: checking the statistical validity of a result – that is, the probability of chance having had an influence upon it. This is usually calculated in relation to an objective of 95% certainty that chance has had no effect (5% level). Significance can also be tested to 1% and 0.1% levels.

sill: an igneous intrusion of relatively hard rock that can produce upstanding *scarp* faces or *waterfalls* where the hard layer is crossed by a river. The sill is a sheet of magma that has been injected along a bedding plane in sedimentary strata. Examples of sills are the Palisades Sill in New Jersey and the Great Whin Sill in northern England.

sima: the rocks in the Earth's crust that consist mainly of silica (si) and magnesium (ma). The density of the sima rocks that largely form the ocean floors is between 2.9 and 3.3 and this layer underlies the *sial.*

single European market: the agreement between the *EU* countries that from January 1, 1993, the trading differences between member countries were to be eliminated so that businesses could treat the whole of the Community as their home market. This was intended to be achieved by the abolition of three trading restrictions: *nontariff barriers*, physical customs controls and different technical standards and taxation levels.

sinuosity: the degree to which a river deviates from a straight line course. The sinuosity ratio is the ratio of the channel length to the straight line distance between the same two points along the center line of the valley. River courses are commonly sinuous or *meandering*.

site: the characteristics of the actual point at which a settlement is located, which would have been of major importance in the initial establishment of that settlement and its subsequent growth.

situation refers to the location of a settlement relative to its surroundings (other settlements, rivers, communication lines and surrounding relief).

skewed result: a *frequency distribution* in which the number of findings is not balanced around the *mean* figure. The findings may be skewed toward values below the mean (*positive skew*) or above the mean (*negative skew*).

slash and burn: see *shifting cultivation*.

sleet is a mixture of *rain* and snowflakes formed when the temperature higher in the atmosphere is below freezing allowing *snow* to form, which partly melts as it falls through the warmer air beneath.

slope development: the result of the interaction of several factors that determine the form of the slope. The major factors are rock structure, *lithology*, climate, soil, vegetation, and the influence of man. All of these factors operate through time, making slope evolution a very complex process. The study of slopes has led to different theories as to how slopes develop through time, the main ones being:

- slope decline – slopes going from the steep to the gentler
- slope replacement – slope goes backward but basal slope becomes gentler
- parallel (slope) retreat – the slope retreats but maintains the same steepness.

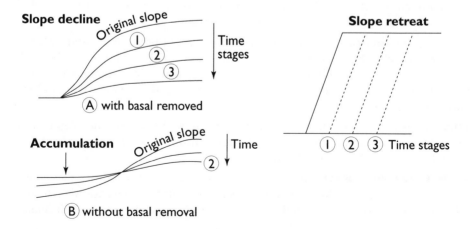

Slope replacement

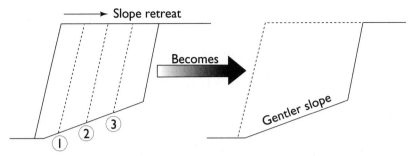

Slope development

slope element: the facets of a slope that go together to give the slope its general character.

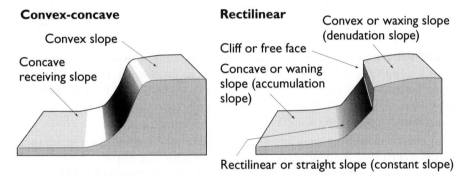

Two slope element models

slumping: a form of mass movement on slopes that involves both sliding and flow. Weathered materials on the slope accumulate and, on reaching a critical mass, move downward in a general curvilinear plane, producing a rotational movement. Some lubrication by water is evident before the movement can take place, unlike a *landslide*, which in general tends to be a dry movement.

smog: a mixture of smoke and *fog* produced through the emission by factories of smoke and chemical *pollutants* added to the output of domestic coalfires. This was a very common occurrence in British cities in the 19th and early 20th centuries, many of the smogs being so thick that the term "pea soup" was applied to them. In December 1952, a smog in London lasted for several days and was responsible, so it was claimed, for over 4000 deaths. The government then brought in legislation in the Clean Air Act of 1956.

More recently, there has been the increase of *photochemical smog*, for which the emissions of car exhausts are seen to be the major culprit. The city of Los Angeles has experienced this phenomenon for a number of years, but photochemical smog occurs over most industrialized and motorized cities when *anticyclonic* conditions prevail and trap the pollutants.

snow: ice crystals aggregating to form a larger unit, a snowflake. The original ice crystals form under the same conditions as rain, except that as the temperatures are below zero, the vapor forms immediately into a solid. The largest falls of snow occur when air temperatures are just below zero. When it is very cold, the air holds very little water vapor; thus the old saying that "it is too cold to snow" has some basis of truth.

social charter is a set of statutory and nonstatutory measures designed to harmonize social legislation within the *European Union,* alongside economic measures. It covers twelve areas: freedom of movement, social protection, vocational training, health and safety, elderly persons, protection of children and adolescents, disabled persons, sexual equality, living and working conditions, employment and remuneration, collective bargaining and the right to strike, information, consultation and participation.

social elite: the social class that has the best living conditions in a population. It may be identified by ascertaining the occupation of the head of the household and/or the level of income of that person. The highest grade band of occupation of head of household is top professional, for example lawyers or directors of large companies. Such people usually have lavish lifestyles and purchasing habits.

social provision refers to the basic needs of housing, education, sanitation and health care that a society ought to provide for its inhabitants. It may also refer to other general needs such as community facilities (for example, libraries, swimming pools and shopping facilities), and environmental services (for example, refuse collection and pollution control).

The nature of, and demand for, social provision may vary in different areas, and at different times. Demographic change in an area will require changes in the expected social provision. For example, a longer life expectancy will result in increased provision of residential care facilities, and greater demands being placed on the healthcare facilities, both locally at daycare centers and more widely in terms of geriatric nursing.

social segregation is the clustering together of people with similar characteristics into separate residential sections in an area. The characteristics may include ethnic and cultural origin, age group, income group, educational background or any other aspect of social class. It is often illustrated by the nature of the housing in an area. Examples could include private housing in comparison with public housing and houses in comparison with apartments.

In recent years, social segregation has become more complex with the process of *gentrification.* Affluent social groups have moved into areas associated with lower income groups and have upgraded the quality of the housing stock. Sometimes this has occurred according to a fashion, whereas in other cases it has happened as part of a *regeneration* policy.

soil acidity: the measure of the concentration of hydrogen ions within a soil. It is measured on the *pH* scale. Soils in upland areas tend to be more acidic as the heavier rainfall leaches out the alkaline substances. If soils are highly acidic then iron and aluminum compounds become mobilized, which can poison plants and organisms.

soil conservation: the maintenance of the ability of a soil to provide an optimum growth of plants. Soil conservation techniques exist to protect a soil from erosion, and to maintain or enhance soil *fertility*. They include:

- the addition of animal and/or chemical *fertilizers* to a soil
- the rotation of crops to ensure that nutrient-demanding crops, for example cereals, are alternated each year with soil-restoring crops, for example ley grass or peas
- *afforestation* on steep slopes to prevent soil from being washed away
- terracing on steep slopes and contour plowing on gentler slopes. By these methods water is given more time to soak into the ground
- the construction of lines of stones across water courses in dry areas so that the limited supplies of water can be trapped
- reduced sizes of herds in semiarid areas to reduce *overgrazing*.

soil erosion: the washing away or blowing away of topsoil such that the fertility of the remaining soil is greatly reduced. Soil erosion is most rapid in those areas where there is the misuse of the land by people. Some activities that cause soil erosion are:

- the removal of vegetation by either chopping down trees or *overgrazing* by animals. In both cases the soil is exposed to the wind and the rain
- the overcultivation of the soil by growing the same crop in the same field year after year (*monoculture*), thus weakening its structure
- the compaction of the soil by the use of heavy machinery. This reduces the rate of *infiltration* into the soil; water flows across its surface and therefore erodes it
- the plowing of land at right angles to the direction of the slope. This encourages *rilling* to take place.

soil fertility: see *fertility (soil)*.

soil moisture graph: illustrates the relationship between *precipitation* and *potential evapotranspiration* for a place over the period of a year. The figure below shows a typical soil moisture graph for a location in West Africa that experiences a tropical wet/dry climate. Precipitation is greater than potential evapotranspiration between June and September, whereas potential evapotranspiration is greater than precipitation between October and May.

- When precipitation is greater than potential evapotranspiration, at first there is some refilling of water into the pores within the dry soil. This is soil moisture recharge. When the soil is saturated, excess water will have difficulty infiltrating into the ground, and may flow over the surface. This is soil moisture surplus
- When potential evapotranspiration is greater than precipitation, water is at first evaporated from the ground surface and transpired from plants. Water may also be brought up to the surface through *capillary action* and then evaporated. This is soil moisture utilization. However, eventually the soil will dry out completely, creating a soil moisture deficit.

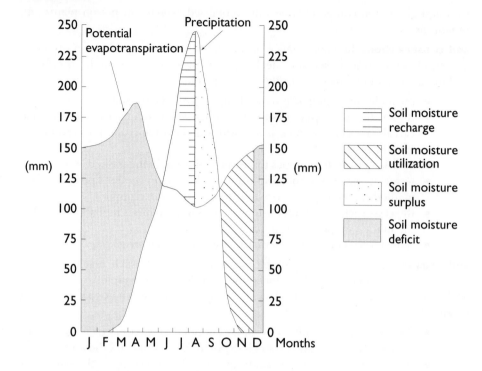

Soil moisture graph – for an area with a tropical wet/dry climate in the northern hemisphere

soil profile shows the variations that occur in the characteristics of a soil vertically down from the ground surface to the underlying *parent material.* A well-developed soil will tend to have a series of horizontal layers within the profile, called horizons, which can be distinguished by their color, texture or chemical composition.

The main soil horizons that occur in most soil profiles are designated by letters:

- O – the uppermost organic horizon comprising of decomposing organic matter
- A – the topsoil, containing both mineral matter and organic matter
- E – the eluviated horizon, which is having nutrients and other substances washed out of it by downward percolating water
- B – the subsoil, generally illuviated, having nutrients and other substances washed in by downward percolating water.

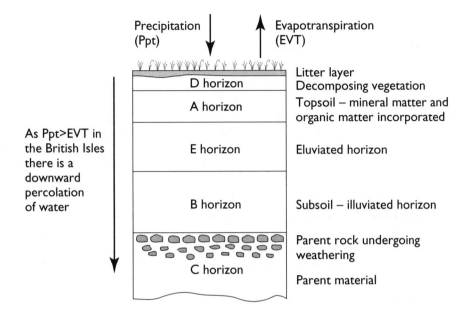

Precipitation (Ppt)

Evapotranspiration (EVT)

As Ppt>EVT in the British Isles there is a downward percolation of water

D horizon	Litter layer Decomposing vegetation
A horizon	Topsoil – mineral matter and organic matter incorporated
E horizon	Eluviated horizon
B horizon	Subsoil – illuviated horizon
	Parent rock undergoing weathering
C horizon	Parent material

The soil profile in the British Isles

soil structure: the manner in which individual particles of soil aggregate together. These aggregates are called *peds,* and they are stuck together by organic matter, and the secretions and mucilages from soil fauna. It is the shape and alignment of the peds that determine the size and number of pore spaces through which water, air and roots can penetrate. They also, therefore, influence the agricultural value of the soil.

There are several types of soil structure, which are not always easy to see in the field. The main ones are:

- crumb peds – small clumps of soil similar to breadcrumbs. They are the most productive, as they are well aerated and well drained
- blocky peds – large brick-like shapes, often enriched with clay
- platy peds – a layered plate-like pattern to the soil, through which water has difficulty draining. It may be created by the use of heavy machinery on a soil that compacts it.

soil survey techniques are used for the collection of data concerning the varying soil types in an area. The following suggestions may be applicable:

- the use of maps – topographic, soil and geological – in order to identify the possible locations of varying soil types
- the use of appropriate equipment – spade, trowel, auger, meter rule, and a camera
- obtaining permission to go onto a landowner's land
- the digging of a pit, or the drilling of an auger, or the use of exposed sections in cuttings or landslips

- the description of the soil in terms of its profile and constituent horizons: the varying textures, structures, colors, acidity levels, organic content, stoniness, land use, vegetation type, etc.
- the returning of the location to its original state – filling in a pit, replacing turf, and expressing thanks to the landowner.

soil texture is the composition of a soil in terms of the varying proportions of different sized mineral particles. Soils consist of particles that are either clay (less than 0.002 mm diameter), silt (between 0.002 and 0.02 mm diameter) or sand (over 0.02 mm diameter). Soils that have a high proportion of sand particles have a sandy texture, and are free-draining and easy to work. Soils with a high proportion of clay particles have a clay texture, and retain water easily. Such soils are heavy, often waterlogged and difficult to work. An ideal soil texture is a loam, which has an even distribution of the three different-sized particles.

solar constant: the amount of *solar energy (insolation)* received per unit area, per unit time on a surface at right angles to the sun's beam at the edge of the Earth's atmosphere. Despite its name it does vary slightly according to sunspot activity, but this is unlikely to influence daily or yearly weather, although it may influence long-term global *climatic change.*

solar energy: see *insolation* and *energy budget (the Earth).*

solifluction: the movement of soil downslope in *periglacial* areas caused by the summer melting of the surface layer (or *active layer*). During the summer the soil above the *permafrost* melts but the water is unable to drain through the soil as it is still frozen underneath. On slopes as low as 2°, this saturated layer can start to move as a *mudflow.* The end product is often a lobe of material at the base of the slope, which may combine with other lobes to give a terrace-like effect.

solution: the removal of dissolved minerals and weathered products by rainfall and percolating groundwater.

solution load: the part of the stream's *load* that is dissolved in the water. Almost all rocks are soluble to some extent, and when streams cross over areas that are susceptible to chemical solution, minerals are dissolved. This is most marked on calcareous rocks such as limestone where carbonic acid dissolves calcium carbonate.

South: refers to the less developed countries of the tropics and subtropics as identified by the *Brandt Commission* (1980) and as such is often regarded as being synonymous with the *Third World.* It includes the areas of South and Central America, Africa and most of South and Southeast Asia. The South faces particular problems in relation to food supply, the need to reduce poverty and the need to establish production systems through investment in development of resources.

sovkhoz: a large *collective farm* in the former USSR that is owned and managed by the state and where the farm workers are state employees rather than shareholders.

spatial displacement: the movement of an activity from one point in the landscape to another. For example, in relation to crime, if one area develops strategies to combat crime such as installing alarms, starting a *Neighborhood Watch* scheme or obtaining more regular police patrols, then crime may well be displaced to another part of the urban area that has not yet developed these responses.

spatial margins to profitability: the boundaries of the area in which a firm can operate and make a profit. This theory takes into account both revenue and costs. Revenue depends on demand, which in turn will be influenced by population density and level of disposable income. Costs include production costs, made up of salaries, rents, capital inputs and transportation costs. The spatial margins are defined by the intersections of the space-cost curves and the space revenue curves. In the example the revenue is constant across the landscape while costs vary. The optimum location is the point of maximum profit, and in this case is also the least cost point. Ma and Mb indicate where profit gives way to loss on this landscape, and they are the margins to profitability. If revenue also varies across the landscape then the point of maximum profit will not coincide with the point of least costs.

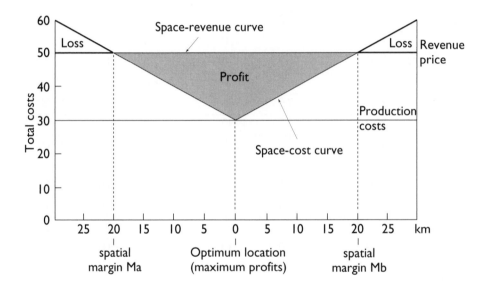

Space-cost/revenue curves

Spearman Rank correlation: a process that produces a numerical value to summarize the relationship between two sets of data. It is based on the ranks of the individual values of the two variables rather than on the values themselves.

The process can be illustrated by using the example of the *correlation* between distance from the sea and annual temperature range for 16 weather stations across North America (see figure below):

- the individual values of the two sets of data are arranged in rank order. The highest value of each variable is given the rank 1 and successive lower values are given ranks 2, 3, 4 and so on
- where two or more of the values are of equal rank, each of the values has been given the average of the ranks that would otherwise have been allocated

- the difference between the two sets of ranks (d) is then calculated for each pair of variables for each station, and this value (d) is then squared (d²)
- the value of d² is then placed into the formula given, and the correlation coefficient is calculated
- the answer for such a calculation will always be between 0 and ±1. A positive value indicates a direct relationship between the two variables, and a negative value indicates an indirect relationship. The nearer to +1 or −1 the stronger the relationship
- the value of the correlation coefficient, however, can only be ascertained by testing its significance. (See *significance testing*.)

Station	Temperature range (°C)	Rank	Distance from sea (km)	Rank	d	d²
1	12.5	16.0	10	15.5	0.5	0.25
2	15.0	15.0	110	13.0	2.0	4.00
3	23.0	12.5	400	9.0	3.5	12.25
4	24.0	10.5	520	7.0	3.5	12.25
5	24.0	10.5	630	6.0	4.5	20.25
6	27.0	9.0	980	4.0	5.0	25.00
7	35.0	3.0	1290	1.0	2.0	4.00
8	38.0	1.0	850	5.0	4.0	16.00
9	32.0	4.5	1230	2.0	2.5	6.25
10	28.0	8.0	1140	3.0	5.0	25.00
11	36.0	2.0	250	11.5	9.5	90.25
12	32.0	4.5	490	8.0	3.5	12.25
13	31.0	6.0	350	10.0	4.0	16.00
14	29.0	7.0	250	11.5	4.5	20.25
15	23.0	12.5	100	14.0	1.5	2.25
16	22.0	14.0	10	15.5	1.5	2.25
					Σd^2	268.50

Spearman Rank correlation coefficient $Rs = 1 - \dfrac{6 \times \Sigma d^2}{n^3 - n}$ where n is the number of pairs of items in the sample.

$$Rs = 1 - \frac{6 \times 268.5}{16^3 - 16}$$

$$= 1 - \frac{1611}{4080}$$

$$= 1 - 0.39$$

$$= +0.61$$

An example of a Spearman Rank correlation

specialization: the division of a work process into separate job functions so that individuals or groups can develop expertise by specializing. Thus people become skillful and productive in particular occupations, and the wider community will benefit. At an individual scale, this is being called into question by the Japanese approach of multiskilling. On a national scale it is still the case that countries specialize in certain agricultural or industrial goods. However, as the *transfer of technology* takes place on an increasing scale, this too will begin to weaken.

species: biological populations in which the individual members can interbreed with each other, but not with other species. Different trees, plants and animals form species.

sphere of influence: the area around a settlement that comes under its economic, social and political influence. This sphere of influence can be of different sizes for each of the different functions provided by the settlement. Similarly, the sphere of influence will tend to be greater for larger settlements than for smaller ones.

It can be identified:

- theoretically by the breakpoint theory of William J. *Reilly*
- by fieldwork methods such as:
  - (a) questionnaire surveys – asking people in an urban center where they have COME FROM to obtain a service or to shop or work; or asking people in rural areas where they GO TO for similar services
  - (b) establishing the areas served by certain functions such as delivery areas for electrical or furniture stores; market areas for real estate agents, local newspapers and hospitals; or the areas served by local transportation providers.

spit: see *sandspit.*

spontaneous settlement: see *squatter settlement.*

spread effect: the transmission of capital, resources and people from the *core* area to the *periphery*. The term, which was coined by Gunnar *Myrdal*, was applied by John *Friedmann* in his core-periphery model. As a core area continues to attract wealth, population and economic growth, it reaches a point where diseconomies of scale may set in. Congestion causes an increase in transportation costs, the price of land rises, which makes expansion costly, pollution costs may increase. Faced with these rising costs, organizations may decide to *decentralize* to the periphery. This will develop a spread effect.

spring: the emergence of underground water at the ground surface. It may take a variety of forms:

- a general seepage resulting in a patch of damp ground
- a weak localized flow of water forming a small stream
- the emergence of a river of some size, called a resurgence.

A spring occurs where a *permeable* rock, for example limestone, overlies an *impermeable* rock, for example clay. Water passing through the permeable rock cannot penetrate downward any further, and so has to reemerge. A spring may also appear where the *water table* reaches the surface.

spring-line: a series of *springs* located along a valley side, usually at the same height. In the case of a *chalk* or *limestone cuesta* (*escarpment*), a spring-line will occur on both the *scarp* and dip slopes. Settlements frequently occur at these springs giving rise to a series of villages spaced regularly along the foot of the slope – spring-line settlements.

spring tide: an exceptionally high tide followed by an exceptionally low tide that occurs shortly after each of the new and full moons of each month. Consequently, there are two spring tides each lunar month, and they are approximately 14 days apart. They are caused by the moon and sun being in alignment and so exerting a greater gravitational pull.

squatter settlements (or shanty towns) are established by people who have occupied land illegally and are using it to build their homes, which are constructed from anything that the squatters can obtain. They are found in and around every major urban center in the *Third World*, and it is not unusual to find that as much as 50% of the city's population live within such a settlement. Different names are used around the world:

- favelas (Brazil)
- barriadas (Latin America)
- bidonville (North Africa)
- bustee (India).

They take many forms and are found on all types of land. In Hong Kong, the lack of land for building has led to the squatters living on junks (boats) in the harbor and other waterways. Squatter settlements are the result of rapid *urbanization* – the large-scale movement of people from rural areas of a country to the cities. Squatters are not "passive" in their use of the urban environment, but are actively seeking to better themselves, either to make improvements onsite or to move to a new area in order to upgrade their quality of life. Many of the residents of such settlements hold jobs within the city, particularly in the city center. In Lima (Peru) for example, to emphasize the positive aspects of many of the newer squatter settlements, the term "pueblos jovenes" (young towns) has been used for them.

stability: the condition whereby a body of air, if forced to rise, for example, over a mountain range, will return to its original position. Throughout the enforced rise, the body of air is always cooler than the air immediately surrounding it. In terms of adiabatic lapse rates, stability occurs when both the *dry adiabatic* and the *saturated adiabatic lapse rates* are greater than the *environmental lapse rate*. Stability may give rise to small cumulus-type clouds, which do not produce precipitation.

stabilization: the methods by which fragile and unstable coastal areas are protected from erosion. In the case of *dunes*, the sand may be stabilized by barriers and fences that act as windbreaks. However, the most common method is the use of vegetation such as *marram grass* and the planting of trees as shelter belts. Another solution is to remove the cause of the destabilization, such as burrowing rabbits or footpath erosion. In tropical areas, the growth of *mangroves* provides an effective protection against storm waves. Their dense root network collects sediment and thereby acts as a stabilizing mechanism.

stack: an isolated pinnacle of rock standing some distance from a cliff coastline. It is a product of the *coastal processes* that at first create a cave, then an arch, the roof of which collapses to leave a stack.

standard deviation is a measure of the distribution of data about a *mean* value. It describes the dispersion of data on either side of a mean value. A low standard deviation indicates that the data set is clustered around the mean value, whereas a high standard deviation indicates that the data is widely spread with significantly higher and lower figures than the mean.

standard error: the potential difference between the *mean* value calculated from a sample of a data set, and the mean value of the total population of the data set. By definition all samples must have a degree of inaccuracy built into them. This inaccuracy is quantified by the standard error. It is usually worked out by calculating the *standard deviation* of the data and then dividing that by the square root of the number of the sample. The standard error can then be used to identify the *confidence levels* of the sample mean.

Ward	Unemployment %	$x - \bar{x}$	$(x - \bar{x})^2$
1	37.2	16.1	259.21
2	36.5	15.4	237.16
3	21.1	0	0
4	33.9	12.8	163.84
5	28.2	7.1	50.41
6	24.6	3.5	12.25
7	12.0	9.1	82.81
8	19.0	2.1	4.41
9	17.4	3.7	13.69
10	22.0	0.9	0.81
11	17.8	3.3	10.89
12	15.0	6.1	37.21
13	7.3	13.8	190.44
14	9.1	12.0	144.00
15	15.6	5.5	30.25
$\Sigma x \quad =$	316.70	$\Sigma(x - \bar{x})^2 \qquad =$	1237.38
$\bar{x} \quad =$	21.11	$\dfrac{\Sigma(x - \bar{x})^2}{n} \qquad =$	82.49
		$\sigma = \dfrac{\Sigma(x - \bar{x})^2}{n} \quad =$	9.08

Mean = 21.11
Median = 19.0
Standard deviation (σ) = 9.08

Standard error $= \dfrac{9.08}{\sqrt{15}}$

$= 2.34.$

Calculation of mean, median, standard deviation and standard error – an exercise based on unemployment levels in different wards of a town

standardization is the production or use of products or components that are so identical as to allow them to be fully interchangeable. The achievement of standardized parts is a necessary condition for efficient *mass production* to take place. It is also a key element in *economies of scale*, as modern firms use the same components in different products to enable longer production runs to take place and save on design costs.

staple food crop is the produce that farmers mainly grow and upon which the local population depends for a major part of their diet. In Southeast Asia, for example, rice is the staple food crop in many areas.

state farming system: where the state owns the land and either leases it to collective groups or develops large state-run farms. This system was typical of the former *USSR*. (See *kolkhoz* and *sovkhoz*.)

state industries: in centrally planned economies (*command economies*) the state owns and directs almost all of the industrial development. In China, for example, from 1950 until the early 1980s, the government dictated the site of each factory, the levels of employment, types and methods of production, wage levels, markets and prices within those markets. In such states, individual *entrepreneurs* were not allowed. In economies such as that of the United Kingdom, the state has, at various times in the 20th century, owned and run certain industries that it has taken over (*nationalization*). In recent years, though, the trend has been to return such industries to the *private sector* (*privatization*).

state intervention: see *interventionist policies*.

statistical population: see *population*.

stemflow is the water that runs down the stems and branches of plants and trees during and after a rainstorm so that it can reach the ground surface. It therefore takes place after *interception* has occurred.

stewardship: the benevolent management of the environment. It involves the careful control of developments on a wide scale, and seeks to maintain the *sustainable development* of resources. It also involves the identification of priorities for the environment and the evolution of strategies to achieve them. Stewardship is a concept that has to be undertaken on a macroscale, across international borders and on a global scale, and herein may be its main weakness.

STOLport: an airport designed to facilitate the use of Short Take-Off and Landing airplanes.

stone circles are associated with *periglacial* areas and consist of boulders arranged in irregular and interconnected polygonal shapes. They are usually less than 10 m across, and fine sands and silts can be found at their center. Their formation is due to the creation of an *ice lens* beneath the surface of the ground. The thickening of the ice at the center of the ice lens causes the ground to heave upward. Stones resting on the ground surface are rolled sideways by the heaving ground. On slopes, the circles become elongated because the stones tend to move down the slope. Eventually the polygons become so elongated that stone stripes are created.

storm beach: the highest part of a sand or shingle beach. It is usually above the influence of the high tide, having been created during periods of strong winds and large waves. Seaward of the storm beach there may be a series of smaller ridges called *berms*.

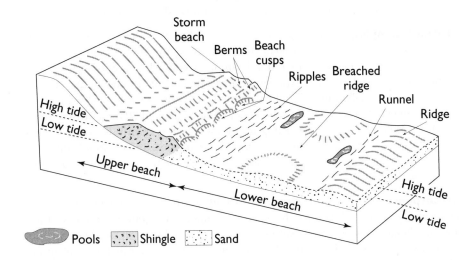

Beach features

storm surge: a rapid rise in sea level in which water is piled up against a coastline to a level far in excess of the normal conditions under high tides. They occur in those parts of the world where, under certain conditions, water can be funneled toward a narrow area of sea.

In the case of the storm surge that affected the North Sea coast of Britain and the Netherlands in 1953, the following events combined to create the crisis:

- an extreme area of low pressure over the North Sea, which caused the sea level to rise
- strong northerly winds blowing down the North Sea producing waves 6 m in height
- high *spring tides*
- rivers already at flood levels flowing into the North Sea.

Over 2000 people were killed by the surge in eastern England and the Netherlands, and this has prompted a number of schemes to prevent such devastation happening again. Similar surges in the Bay of Bengal have resulted in thousands of deaths in Bangladesh.

storm water disposal: the deliberate removal of flood water from an urban area through a system of undersurface drainage designed to discharge water as quickly and efficiently as possible. One of the consequences of rapid removal of water into neighboring stream channels is a "flashy" *hydrograph* with a steep rising limb, a high peak and a shortened lag time. Channels fed by urban catchments respond quickly to rainfall input; this can cause flooding problems down-river of the urban area.

Stouffer, Samuel Andrew: proposed the theory of *intervening opportunities* in 1940.

Strahler, Arthur Newell: devised a method of *stream ordering* as a means of investigating relationships between geometric properties of a drainage network such as the number and length of streams, drainage density and drainage area.

stratified sampling: samples are selected according to some known background characteristic in the *statistical population*. In studying the distribution of land use types in relation to geology, a stratified sample would select points in proportion to the area covered by each type of geology. If 30% of the area was clay, 25% sandstone, 30% chalk and 15% alluvium then for a total sample of 200 points, 60 would be selected on clay, 50 on sandstone, 60 on chalk and 30 on alluvium. In this way the sample is stratified according to a known factor. In investigating the opinions of the local population on some development, the sample should reflect the various interest groups fairly, i.e. it should account for the sex-age distribution in the population.

stratosphere: a layer in the *atmosphere* extending from the *tropopause* up to about 50 km above the Earth's surface. In this layer temperatures generally increase gradually with height. The layer is extremely dry, with no clouds or weather extending above the tropopause, although it does contain the *ozone layer*, which is concentrated at a height of 25 km (90% of ozone is found below 35 km). The absorption of ultraviolet radiation leads to a warming of the stratosphere with maximum temperatures occurring at the 50 km level.

stratus: a type of cloud that is layered and often unbroken. It develops mainly in the lower levels of the *atmosphere* below 2500 m and it forms under stable conditions to produce low grey cloud. This tends to give drizzle but little rain. Higher level stratus is termed altostratus.

stream ordering system: a method of classifying the parts of a stream network. A number of systems have been proposed by Robert E. Horton, Arthur *Strahler* and Forrest Shreve. Strahler's method is most commonly used. In this the smallest streams, which have no tributaries, are termed 1st order streams. Where two 1st order streams join, they produce a 2nd order stream.

Where two 2nd order streams join they produce a 3rd order stream and so on. The order only changes when two streams of the same order join, i.e. if a 2nd order stream is joined by a tributary 1st order stream then it remains a 2nd order. Once the order of streams in the network has been designated the basin can be analyzed. The number of streams in each order can be counted and a graph drawn to show the relationship between stream order and frequency. This indicates the *bifurcation ratio*.

An analysis of this kind allows objective comparisons to be made between drainage basins of different size.

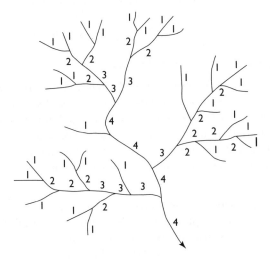

Stream ordering (Strahler)

strip farming: cultivation of a field in long narrow strips as part of a medieval two- or three-field system. This was the basis of the *open field system*, which produced a ridge and furrow landscape.

strip mall: a shopping development usually located on the edge of an urban area, an "out-of-town center." It has a similar background and location factors as a shopping *mall*, and it is smaller than a regional shopping center. Whereas a shopping mall tends to have all the different outlets under one roof, a strip mall consists of a number of independent companies in individual buildings served by one large parking lot. There is a considerable amount of shopper movement from one store to another, many of the goods on offer are complementary and customers may be purchasing a variety of goods. Such developments can be seen on the edge of most large urban areas that provide a large potential market.

structural unemployment occurs when there is a change in demand or technology that causes long-term unemployment. Very often this occurs in particular regions that have been heavily dependent on certain industries, such as coal mining in West Virginia and car manufacturing in Detroit. There is little that governments can do to alleviate such unemployment other than offering retraining for those laid off.

subaerial processes: a term that covers all the physical processes taking place on the surface of the Earth, such as subaerial *weathering*.

subassembly means putting together a component that will later be part of the final assembly of the finished product. On a car *production line*, the assembly of the gearbox would be termed a subassembly.

subcontracting: finding a supplier to manufacture part or all of a product. The main circumstances in which this would be used are:

- when a firm is already operating at maximum capacity and therefore cannot meet further demand in any other way
- when there are elements of a product that the firm is ill-equipped to manufacture efficiently
- when coping with seasonal peaks in demand.

subduction occurs when an oceanic crust sinks below either an oceanic or a continental plate along a convergent or *destructive plate margin*. If subduction is beneath oceanic crust then *ocean trenches* and *island arcs* are formed; if it is beneath continental crust then a line of fold mountains, with associated volcanic activity, forms at the edge of the overriding plate. If subduction is so extreme that it closes an ocean so that two continental plates collide, then complex mountain building occurs, as in the Himalayas.

As the lithosphere sinks beneath the overriding plate it is colder and more brittle than the *asthenosphere*, and this produces stresses. The sudden failure of the lithosphere triggers *earthquakes*, which are concentrated along the line of the subducting plate in a zone called the *Benioff Zone*. When the lithosphere has subducted to depths between 80 and 100 km, and temperatures of 1400°C, the rocks begin to soften and melt, and this less dense material migrates toward the Earth's surface as plutons that produce intrusive and extrusive magma.

subglacial: features that lie at the base of, or beneath, a glacier or ice sheet. Subglacial streams flow in tunnels at the base of the ice and the material deposited along their channels is let down during *deglaciation* to produce *eskers*.

sublimation involves a direct change of state from a solid to a gas without passing through the liquid stage.

submergent features are produced along a coastline that has experienced a relative rise of sea level, as in the postglacial period. On upland coasts features such as *rias* and *fjords* are formed when former valleys are inundated, while in lowland areas broad shallow estuaries are produced. A rise in sea level also causes *aggradation* of the lower river valley, which produces an extensive area of easily flooded marsh and mud flats.

subsidence: generally means sinking to a lower level. When applied to land movements it often occurs when underground mining has weakened the subsurface structure, which results in the surface layer sinking. In meteorological terms, air that descends is subsiding. In *anticyclonic systems* subsidence leads to a warming and drying of the air; stable air tends to subside and bring calmer weather. The subsidence of cold air from valley sides into the valley floor can create a *temperature inversion*.

subsidy: a type of government intervention that takes the form of financial grants given to an industry or group of people to encourage production of a particular commodity or to generate greater output. It may be directed toward consumers in terms of reducing food prices. Subsidies are often provided in agriculture to raise incomes or through the *CAP* to promote production of crops.

subsistence: a type of agriculture in which the produce is consumed mainly by the farmer and the family who work the land or tend livestock. There is little surplus to be sold or traded. Farming may be *intensive* or *extensive* in operation and may focus on crop or livestock production. Crop systems include *shifting cultivation*, rice production (Southeast Asia), and bush fallowing. Subsistence livestock farming is basically pastoral nomadism.

suburbanization is the outward growth of urban development to engulf surrounding villages and rural areas. As cities increased in population in the 19th and early 20th centuries, the physical expansion of the urban area also occurred and neighboring rural areas were taken over by this outward expansion. This was facilitated by the growth of public transportation systems and the inhabitants of these areas were able to travel to work in the urban areas as commuters. Initial growth was linear along the railroad lines but areas became in-filled as tram, bus, underground systems and later the private car gave individuals greater freedom of movement. The process of suburbanization may also be viewed as the suburbanization of settlements physically separated from the urban area (sometimes called "extended suburbanization"). This represents a major trend in the redistribution of population away from city centers.

succession: a series of changes that take place to a plant community. This may be a *primary succession*, or *prisere*, if the development begins on a surface that has previously not been vegetated, or it is termed a *secondary succession*, or subsere, if the original vegetation has been cleared or destroyed naturally.

succulent: a plant that is able to store water in its stem or leaves as an adaptation against drought conditions. Species of cacti have no true leaves, but are formed of thick green stems containing large water-storing cells that absorb large amounts of water and swell up. The lack of leaves and the thick waxy covering to the stem helps to reduce transpiration. The baobab of the savanna regions has a huge trunk, up to 9 m in diameter, which has a soft and spongy interior allowing storage of water. The giant euphorbia, which has a stem similar to cacti, is also common in the thorn scrub of East and Central Africa.

sunrise industry: one positioned in a rapidly growing market, usually based on new technology and innovation.

sunset industry: one believed to be in terminal decline, with obsolete technology and an obsolete product.

superimposed drainage is a system established on a series of younger rocks, lying on a markedly different older series. The drainage will become adjusted to the structural pattern of the younger series. In the course of time, erosion will remove the younger layers, leaving the established drainage pattern cutting down into the older rocks, the structure of which bears no relation to the drainage system.

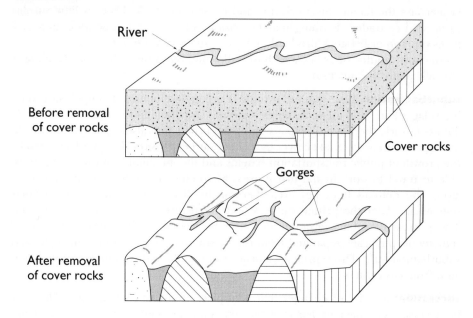

River

Before removal
of cover rocks

Cover rocks

Gorges

After removal
of cover rocks

Superimposed drainage

superstore: a large *retail* outlet with substantial parking space that offers a very wide range of goods within a self-service store.

supraglacial: that which is happening or to be found on the surface of the glacier.

suprastate: the grouping of countries into a much larger organization. The *European Union* is an example of such a suprastate, which by 1995 consisted of 15 member countries.

surf zone: the area on the foreshore of a beach between the breakpoint position of the approaching waves and the upbeach (landward) limit of the *swash*. The extent of this area depends on the type of wave, *constructive* or *destructive*, and the angle and nature of the beach material. A steep beach with large sediment quickly absorbs the swash energy that restricts the zone. A shallow angle beach of fine material allows the swash to extend a greater distance up the shore.

suspended load: the material small enough to be carried along in the flow of water in a stream. Some particles, such as silt or clay, can be carried by turbulence within the water; the greater the turbulence the larger the size of the particles that can be carried. The relationship between stream velocity and particle size (caliber) erosion, transport and deposition has been determined experimentally by Filip *Hjulstrom.* The greater the velocity, the greater the size and the amount of material that can be carried in suspension.

sustainable agriculture: a system by which food production can expand but in a way that does not destroy the natural environment or put an excessive strain on resources. Many believe that sustainable strategies in agriculture are the only way forward if all societies are to obtain a more secure food environment. Sustainable agriculture is based on the following:

- replenishment of soil *nutrients* as much as possible with organic material
- maintenance of the soil's physical structure
- minimal off-farm environmental contamination
- maintenance of habitats for pollinators and *biological pest control* agents
- conservation of genetic resources of crop and animal species farmed
- direction of technology to change away from the use of *nonrenewable resources* and subsidized energy toward *renewables*
- continual cover of the soil by vegetation
- no increase in soil toxicities
- agriculture to be profitable enough to secure adequate subsistence and income for the farmer's family.

sustainable development: defined as "development that meets the needs of the present, without compromising the ability of future generations to meet their own needs." The environment should be seen as an asset, a stock of available wealth, but if the present generation spends this wealth without investment for the future, the world will run out of resources. If, however, we use this capital to research and develop new resources for the future, we can build machines that will substitute for the environmental resource. A good example is the construction of solar panels to replace oil and coal. Examples of how this can be put into practice are already being seen:

- fitting cars with catalytic converters
- replacing *fossil fuels* in power generation with *renewables* such as *wind, hydroelectric, solar* and *tidal power.*

sustainable resources: resources such as soil, forests and fish stocks. If these are well managed, by replanting or manuring for example, the supplies can be sustained. For each there is a maximum sustainable yield, beyond which extraction rates will exceed renewal rates.

swash: the body of water rushing up a beach after a wave has broken. It may carry material up the beach away from the sea.

swiss cheese effect: a distribution across an area that is patchy, the patchiness being described as the "swiss cheese effect," i.e. there are holes in it.

synergy occurs when the whole is greater than the sum of the parts, i.e. when $2 + 2 = 5$. On some *science parks*, for example, there can be an intense localized interaction between different firms and *entrepreneurs*, research institutions, local banks and business service organizations. The benefits of these links and relations generates outputs, the sum of which is greater than the sum of the individual parts.

synoptic chart: a map that shows the weather for a particular area at one specific time.

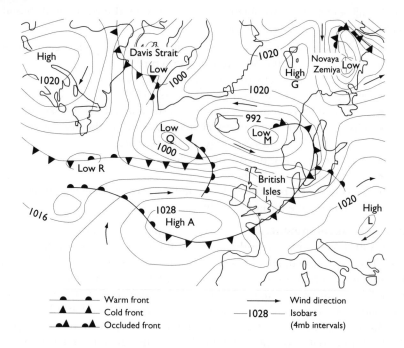

A synoptic chart

system: any set of interrelated components or objects that are connected together to form a working unit or unified whole. In geography it is usual to recognize two general types of systems:

- closed – there is transfer of energy, but not matter, between the system and its surrounds. The Earth is an example of such a system
- open – systems receive *inputs* and transfer outputs of energy or matter across the boundaries between them. Most natural systems are open ones.

systematic sampling: a method in which the sample is taken in a regular way, i.e. every tenth house, every tenth person, at grid intersections in an areal sampling exercise.

systemization was the name given to the controversial program of President Ceausescu in Romania, where hundreds of rural villages were to be destroyed and replaced by urban-style "agroindustrial complexes." The aims were:

- to create more agricultural land
- to "eliminate the differences between the towns and the villages"
- to remove the only means of subsisting independently within the socialist state.

The program had only affected around ten villages when Ceausescu was overthrown in December 1989.

T

taiga is a term, originally from Russia, for the coniferous, or *Boreal*, forest that extends across the northern parts of Russia, Scandinavia and Canada.

talus is an accumulation of weathered fragments on a slope. This may be a thin layer of debris resulting from *weathering* of the underlying rock or it may be a thicker layer resulting from the movement of rock fragments downslope from a rock face. In the latter case it forms a *scree* at the base of a cliff (free) face. The talus may form in cones and is likely to move downhill due to gravity and expansion and contraction; this is termed talus creep.

tariff: a duty or tax imposed by one country on the imports from other countries. This may be used as a device to protect home-based manufacturers from foreign competition, particularly when overseas production is cheaper as a result of lower raw material and labor costs.

technological treadmill: this refers to agriculture in the *First World* where small farmers are stuck borrowing money to pay for technology in order to keep up with larger-scale farms, particularly *agribusiness*. Farmers, because of a price-cost squeeze, are trying to maintain their income levels by raising output, usually by intensification or specialization using *hi-tech* imputs. This technology is expensive, therefore further increases in output are required to pay for it, which could lead to *overproduction* and a fall in farm prices and thus incomes. The only way for the farmer to keep going in these circumstances is to buy more technology or more land. Either way, the farmer appears to be stuck on a treadmill that they cannot get off if they wish to compete.

tectonic processes cause movement within the Earth's crust that results in uplift or depression, *folding, faulting,* warping or plate movement at convergent or divergent boundaries. The landforms produced by these processes, such as rifts, block mountains, fold systems and escarpments, are termed tectonic relief.

temperate: a term used to describe conditions that are not extreme, as in a temperate climate, where seasonal temperatures do not display a wide range. The temperate latitudes extend from the tropics to the Arctic and Antarctic circle.

temperate grasslands: extensive areas of natural grassland situated in the dry continental interiors of North America (prairies) and Russia (steppes). They are found in temperate latitudes where the annual rainfall is between 250 and 750 mm. They are also found on continental east coasts between 30° and 40° south of the equator in the Pampas of Argentina and the Canterbury Plains of New Zealand. The Veld of South Africa and the Murray-Darling Basin of Australia are more continental in location. Although grasses grow in a wide range of climates, they only become dominant, as in the temperate grasslands, where trees or shrubs are unable to grow.

The transition from grassland to forest in the cooler northern areas takes place at

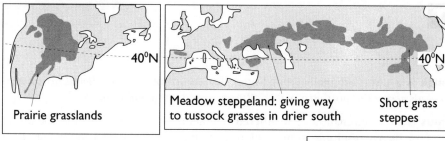

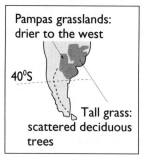

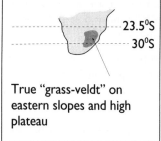

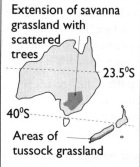

Temperate grasslands

about the 500 mm annual rainfall isohyet. Further south, in the warm temperate areas the transition occurs around 750 mm isohyet. These grasslands have probably been influenced by fire and human exploitation. In the prairies, the rainfall decreases from east to west and there is a change from tall grass to short grass prairie; there is a corresponding change in soil type with *chernozem* in the drier east and prairie soils in the area of taller grass to the west. Within these soils the spring *leaching* and summer *capillary action* (when higher temperatures produce an evaporation demand that exceeds the summer maximum of rainfall) create a layer of calcium carbonate nodules in the soil. This layer develops at shallower depths westward as rainfall decreases. The dominants are grama and buffalo grass; the deep roots may extend down 2 meters to the water table, binding the soil together. The soil is the major nutrient store as the decay of grasses in the summer, together with high temperatures that favor breakdown, provide a rapid accumulation of *humus* in the soil.

temperature inversion: an atmospheric condition in which temperature increases with height rather than the more usual relationship shown by the *environmental lapse rate*. A number of situations can produce inversions:

- on a clear night when heat is radiated from the Earth's surface, the air near to the surface is cooled by conduction of heat to the cold ground. The lower layer of the atmosphere is therefore cooler than the air aloft
- at night, the colder denser air on the upper slopes of a valley side descends into the valley bottom, displacing warmer air aloft. This produces a *katabatic wind* and can develop "frost hollows" on valley floors, which can affect sensitive crops

- when a warm air mass passes over a colder land surface it will be cooled from below, and the normal temperature profile will be altered
- when air is subsiding in an *anticyclone*, or at a kata-front in a depression, the air is warmed *adiabatically* and is warmer than air at ground level.

Inversions over urban areas can help to trap *particulates* and pollutants in the lower layer of the atmosphere because *stability* increases at the level of the inversion.

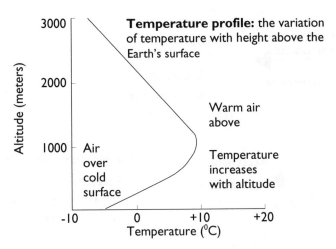

Temperature inversion

temperature profile: the variation of temperature with height above the Earth's surface.

tenure: the means by which property, such as land and buildings, is held by an individual or organization.

In terms of housing, tenure may be by owner-occupancy or tenancy. Tenancy may take a variety of forms: to a private landlord, to the government or to a cooperative.

There are also a variety of forms of land tenure:

- freehold ownership – where the land is owned by an individual
- cash tenancy – where a fixed cash rent is paid by the tenant to a landowner
- *share-cropping* – where a tenant gives the owner an agreed share of the produce
- state ownership – where the land is the property of the state and farmers act as tenants of the state. (See *collective farming, kolkhoz, sovkhoz* and *state farming systems.*)

tephra: the ash and pumice that is ejected during a volcanic eruption and then deposited on a landscape. (See also *pyroclastics.*)

terminal moraine: a low crescent-shaped ridge stretching across the width of a valley floor that marks the furthermost point reached by a *glacier*. It is composed of an unsorted and unstratified assemblage of debris that has been brought down the valley by a glacier. The snout of the glacier remained stationary for a long period of time, representing a point when *ablation* and *accumulation* were equal.

During this time *supraglacial, englacial* and *subglacial* materials were deposited at the snout as *meltwater* flowed away.

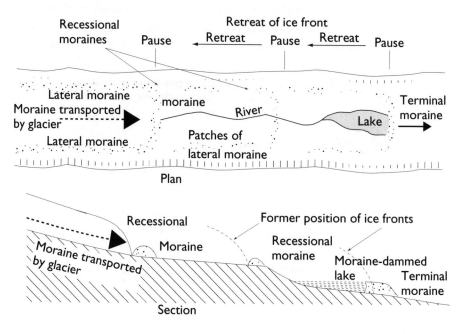

Terminal, recessional and lateral moraines

If a glacier retreats intermittently, with several such pauses, then a series of moraines across the valley floor are created, each marking the position of the ice front. These moraines are similar in nature to terminal moraines, but are called recessional moraines.

terrace: a raised level area of ground on a steep hillside. Terracing helps to slow down the rate of surface run-off of water, giving it more time to infiltrate.

Terraces are common features in areas of intensive rice farming, where they act as a means to reduce *soil erosion*, and to make profitable use of slopes that would otherwise be difficult to cultivate. (See also *river terrace*.)

terra rossa: a red-colored soil developed in areas with a calcium carbonate parent material. It occurs in areas that have high levels of seasonal precipitation and high temperatures. The parent material is greatly weathered, and silicates are leached out of the soil to leave residual iron-rich deposits. It tends to accumulate in depressions within the limestone, and where the vegetation type is *garrigue*. (See also *calcareous soils*.)

terrorism: the use of fear-creating methods to either govern or coerce a government or society into changing its attitude to an issue. Terrorism often occurs in those areas where the desire for *separatism* is strong. Examples include Northern Ireland prior to 1994, and the Basque region of France and Spain. Terrorist acts frequently take place in large cities in order to gain maximum publicity and disruption.

tertiary sector industries provide a service. These include employment in education, health, the police and fire services, public transportation, retailing, local government, banking and finance, and the armed forces. It is the tertiary sector that employs the most people in an *economically more developed country* such as the United States. The tertiary sector is also known as the *service sector*.

textiles are materials and fabrics that are suitable for weaving. They may be divided into natural fabrics such as wool, cotton and silk, and artificial fabrics such as nylon, acrylic and other synthetic fibres.

thermal electric power (TEP): electricity produced by means of steam turbines. The heat for steam production comes from the burning of a fuel, such as coal, lignite, oil, natural gas and peat.

thermokarst is the name given to the landscape that results from *periglacial* conditions where the ground surface is very irregular and hummocky with marshy or lake-filled hollows. The major cause is the melting of masses of ground ice, which results in parts of the surface subsiding. Thermokarst in its natural state is often due to changing climatic conditions, but increasingly one of the major causes has been man's development of such areas.

Thiessen polygon: a statistical technique that is used to convert data measured at specific points into an average value for the whole area. In calculating average rainfall within a drainage basin accurate values of rainfall can only be measured where *rain gauges* are sited. This measurement is effectively a point sample. If several gauges are used across the basin, then polygons can be constructed as a basis for calculating an average:

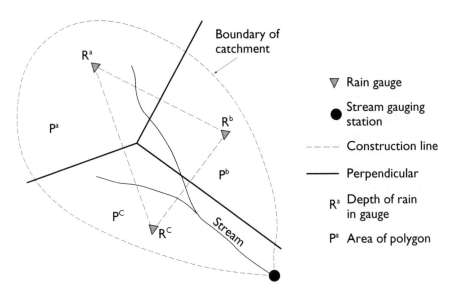

Thiessen polygon

- the location of the rain gauges within the basin are plotted on a large-scale map
- a straight line is drawn from each gauge to its immediate neighbor
- midway between two gauges a line is constructed at right angles to the line joining the gauges
- these lines are extended until they intersect with others to form polygons.

To calculate the average rainfall for the basin:

- calculate the area of each polygon, either by drawing the outline on graph paper or by using transparent graph paper and counting squares
- sum the area of the polygons to give the total area of the basin.

FORMULA   Average rainfall for the basin =

$$\left[R^a \times \frac{P^a}{T}\right] + \left[R^b \times \frac{P^b}{T}\right] + \left[R^c \times \frac{P^c}{T}\right] \text{ etc. for all polygons}$$

where R^a, R^b, R^c are the rainfall figures for gauges a, b, c
P^a, P^b, P^c are the areas of each polygon surrounding gauges a, b, c
T is the total area of the drainage basin, i.e. $P^a + P^b + P^c$.

Third World: the name given to those countries that are also described as *developing, economically less developed countries (ELDCs)* or the *South*. It covers a range of countries from the relatively rich *NICs* to the least developed (*LDCs*).

threshold: the minimum number of customers needed to support a good or a service in *central place theory*. It represents the minimum trade area required to provide sufficient sales to cover the costs incurred by the supplier of the service, and therefore shops offering a particular good must be at least far enough apart to satisfy the threshold. This determines, in theory, the spacing between shops offering goods of different *order* within the shopping hierarchy.

It is also the term applied to the shallow area at the mouth of a fjord where the erosive power of the glacier was reduced by melting. The threshold may be a rock bar, or it may have some depositional material above solid rock.

throughfall is the water that drips off the leaves of trees during a rainstorm. It occurs when the amount of rain that falls on the interception layer of the tree canopy has exceeded the capacity of the leaves to hold water.

throughflow is the water that moves downslope through the subsoil. It is particularly effective where further downward percolation or *infiltration* is prevented by underlying *impermeable* rock. In the *hydrological cycle* throughflow transfers water from the soil storage zone to the channel at a much slower rate than *overland flow*.

thunderstorm: a violent and heavy form of precipitation with associated thunder and lightning. They are produced by convectional uplift under conditions of extreme *instability*, which may develop cumulus and cumulo-nimbus cloud up to the height of the *tropopause* where the inversion produces stability. This causes the cloud to spread out to form anvil clouds. The updraft through the central area of the towering cumulus system causes rapid cooling and *condensation*, leading to the formation of water droplets, hail, ice and supercooled water, which coalesce during collisions in the air. During condensation, *latent heat* is released, which further fuels the convectional uplift. As raindrops are split in the updraft, positive electrical charge builds up in the cloud. When the charge is high enough to overcome resistance in the cloud, or in the atmosphere, a discharge occurs to areas of negative charge in the cloud or to Earth, producing lightning. The extreme temperatures generated cause a rapid expansion of the air, which develops a shock wave that is heard as thunder.

tidal energy: a *renewable* method of producing energy by using the movements of the *tide*. Schemes could, with reversible blades, harness both the incoming and outgoing tides and therefore maximize the use of the site. Places with the maximum *tidal range* offer the greatest potential. The major drawback is cost, in both economic and environmental terms, and this may explain why only two sites are at present in operation: the Rance Estuary in northwest France and on the Bay of Funday in eastern Canada. The arguments for and against tidal schemes include:

Pros:
- renewable
- very reliable and predictable
- large size
- nonpollutant
- also gain benefits of coastal protection of estuary

Cons:
- flooding of wetlands bordering estuaries, often the home of many species of birds, particularly those that are migratory
- could have an adverse effect on spawning fish
- construction cost is very high.

tidal range is the difference between the height of the water at high and low tide. This obviously varies, but a mean figure is normally given for each coastal location, the largest ranges being over 10 meters, e.g. Bay of Funday (Canada) tidal range of 15 meters. Tidal ranges are at their highest at spring tides.

tide: the periodic rise and fall in the level of the water in the oceans and seas, the result of the gravitational attraction of the sun and moon. The moon has the

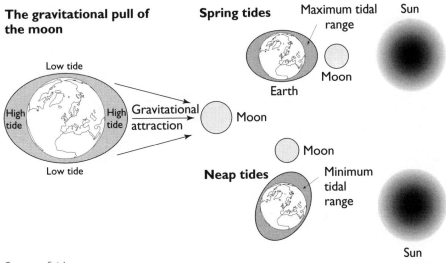

Causes of tides

greatest influence, pulling water toward it to create high tide, with a compensatory bulge on the opposite side of the Earth. In the intervening areas, the tide is at its lowest. Tides run in cycles, following the 28-day lunar cycle; therefore every 28 days the gravitational pull is greatest and this gives the highest tides, known as spring tides. At the point in the cycle 14 days after this event, tides are at their lowest before they begin to rise again. This lowest point in the cycle is known as the neap tide.

till: a general term that covers all the materials deposited directly by ice. They are unsorted and include clays, sands and rocks. The general term for such glacial deposits was once *boulder clay*, but since many contain neither boulders nor clay, the term till is now in widespread use. Till and *fluvioglacial material* collectively make up glacial *drift*, the term that covers all glacial deposits.

TNC: see *transnational*.

Todaro model: an alternative explanation for rural-urban *migration* in the *Third World*. Ideas of such movements have in the past focused on the *perceptions* of individuals in that they fall for the deceptive attraction of the "bright lights" of the city. Michael Todaro argues that people have expectations about the city and these expectations are economically realistic and rational. People move to the city to find better jobs but are prepared to gamble on the lottery of finding work, as urban employment pays far more than rural work. Therefore, even if the chance of finding work is only 50 : 50, given the difference in wage rates, this is a gamble worth taking.

tombolo: a coastal feature, produced by the deposition of sand and shingle, that joins the mainland to an island, for example at Marblehead, Massachusetts. It may be formed by the extension of a *spit* due to *longshore drift* or by an offshore bar pushing onshore. Chesil Beach in England is a spectacular example, extending from West Bay near Bridport, on the coast of Dorset, 25 km southeastward to the Isle of Portland, attaining a height of 14 m above sea level at the southern end.

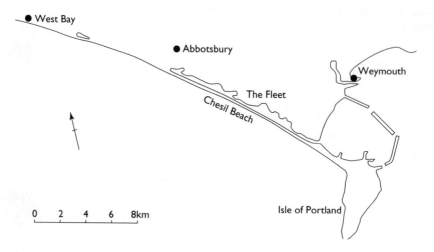

Tombolo

topographical map: a map that shows the surface features of an area to scale. This includes both the physical elements, such as relief and drainage, and human features, such as settlements and communications.

topological map: a map that represents a spatial pattern of places and linkages by means of a diagram or graph. Dots or vertices represent places, while straight lines or edges represent linkages between places. The original scale, orientation and true distance between places is distorted; the map is drawn to show places that are connected and intersections in the system. One famous English topological map is that showing the London Underground system. There is no scale, and stations are not shown in their true location, but the map is very effective for its purpose – it allows travelers to move efficiently around the system because it accurately portrays the links and interconnections. Topological transformations are produced to allow comparative studies of the *connectivity*, or degree of interconnection, between networks and the *accessibility* of individual points in a system.

toponymy is the study of place names particularly in terms of their cultural and environmental derivation. Settlements that originated in the Roman, Anglo-Saxon and Scandinavian periods have characteristic place name endings that often relate to the physical setting or stage of forest clearance. This allows greater analysis of settlement distribution patterns.

topset beds are layers of sediment that are laid down at the landward edge of a *delta*. They consist of coarse sands and silts deposited horizontally as the velocity of the river reduces on entering the lake or sea.

tor: the most distinctive feature of granite landscapes, consisting of an isolated exposure of much jointed rock. Their origin has been the subject of much controversy, with two main theories as to their formation:

- blocks of exposed granite were broken up by *frost shattering* during *periglacial* times. The weathered material was moved downhill by *solifluction* to leave the more resistant rock exposed as the tor

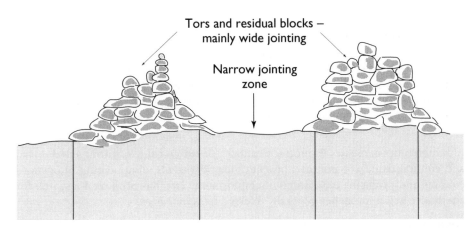

Tors and residual blocks – mainly wide jointing

Narrow jointing zone

Tor formation

- *joints* in granite were widened by chemical action under the surface. Deep weathering then occurred during the warmer periods (*interglacials*) when rainwater penetrated the still buried granite. As the joints widened, roughly rectangular blocks were formed. During colder times, solifluction removed the weathered material to leave the granite outcrops. The joint pattern is important, areas of granite where there were few joints being left as the upstanding tor. This is the more widely accepted theory.

tornado: an extremely violent and destructive weather system characterized by very powerful swirling winds that rotate counterclockwise toward the intense low pressure at the center with speeds up to 300 km per hour. Their development is linked to extreme *instability* in the atmosphere, *convergence* and vigorous updrafts in the air. They are small in extent; the diameter of the storm center may be 100 meters or less and they may last in an area for only a matter of minutes. The life of the system may only last for a few hours. However, they cause an immense amount of damage along their storm path due to wind speed and rapid fall in pressure. This can cause such rapid pressure differences that buildings explode and debris is thrown around in the system. The strong updrafts can uproot trees, lift people, cars and even train locomotives. The USA experiences more tornadoes than anywhere else; on average there are 100 deaths per year and damage running into hundreds of millions of dollars. The main area of development is in the Great Plains area of the Midwest, the Gulf Coast and the Mississippi Basin, an area referred to as "tornado alley." In this area cool air from the continental interior to the north meets warm, moist air from the Gulf of Mexico; this is particularly effective in spring and early summer, most tornados occurring in May and June, when there is also strong heating of the ground surface that adds to the uplift in the system.

tourism: any *recreational* or leisure time activity that involves an overnight absence from the normal place of residence. The development of tourism can generate employment both directly, in the creation of jobs in the hotel and catering trades, and indirectly, through tourist expenditure on goods and services in the area. Transportation, construction and public utilities also benefit through the economic expansion linked to growth. Although the popularity of particular areas may fluctuate, making tourism a less reliable base for economic advancement, this sector is rapidly expanding and it receives considerable private and public investment in some areas. The purchasing power of tourists can be a significant contribution to the local economy, representing a source of foreign earnings (invisible earnings) and as such it is an export activity. The World Travel and Tourism Council (WTTC) claims that this is the world's largest and fastest growing industry, and that the current 500 million tourist travelers will increase by 80% in the next 15 years.

Although tourism creates economic wealth it does often bring environmental, social and cultural damage – polluted beaches, degraded reefs, disappearing indigenous cultures and problems of seasonal unemployment. This has prompted a search for alternative new approaches generally referred to as *ecotourism*.

toxic waste: a source of pollution resulting from the use of metals that produce a poisonous residue or give off dangerous fumes. Many industrial processes release toxic metals such as lead, mercury, zinc, cadmium or discharge sulfuric acid and cyanide. Some chemical pollutants remain toxic for many years and may become concentrated in the *food chain*. Toxic material may be airborne pollution; mercury may be released into the air along with lead from vehicles that are driven by gas combustion. These metals are carried by the wind over long distances and are eventually washed out by rainfall to contaminate the soil. Toxic waste dumping, even at official refuse tips and sites, presents a hazard where leaching carries waste into streams, rivers and *groundwater* supplies. Discharge of pollutants into sea areas contaminates marine shellfish and fish stocks can be influenced; the concentration of the polluting metal tends to become greater at successive *trophic levels*.

trade: the movement of goods from producers to consumers. The basis of this transfer of goods can be explained by the theory of comparative advantage; countries specialize in activities for which they are best equipped in terms of *resources* and technology. Surpluses can then be traded by a country to provide the income needed to buy in goods that cannot be produced efficiently, or at all, in the home economy. Trade is partly in the form of raw materials, foods, beverages, energy supplies and manufactured goods that are termed visible imports or exports. Trade also takes place in services such as finance and *tourism*, which are forms of invisible earnings or expenditure. World trade is dominated by *OECD* countries, i.e. the West and Japan, where much of the trade is between the members of this group. These are all economically advanced states and much of the trade is in manufactured goods. There is also a significant movement of fuel oil from the *OPEC* countries, particularly the major exporters in the Middle East, to the advanced economies and to developing world states, which are deficient in natural energy resources. Raw materials (ores and industrial crops) and foods and beverages from the economically less developed areas have traditionally been traded with advanced nations in return for manufactured goods. This is still important, although as countries have begun to industrialize, they have tended to progress from industries that are "export-oriented," such as basic processing of raw materials, to those that are "import substitution," such as the manufacture of textiles, clothing, shoes, cigarettes and drinks. This change has led to a decline in the return movement of manufactured items.

trade (trading) bloc: a group of countries that share trade agreements with each other, but with *tariff* walls that discourage imports from countries outside the bloc. The *European Union (EU)* is a good example.

trade cycle: see *business cycle*.

trade war: a protectionist battle between governments in which *tariff* barriers against a country's imports lead to retaliation. If a trade war becomes serious enough it can cause a widespread downturn in world trade.

trade winds: the tropical easterly winds that blow in from either side of the equator toward the *intertropical convergence zone* from the subtropical high pressure cells. They form the surface component of the *Hadley cells*.

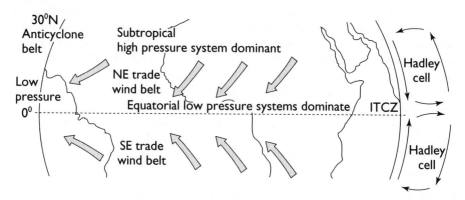

Trade wind belts

transect: an illustrative device for showing features along a line, drawn to indicate how those features change along the line as distance increases. A typical use would be in urban geography, where transects can be drawn from the city center to the outskirts along main roads to show the urban land use changes that occur with increasing distance from that center. Another use would be to show the changing vegetation of a *psammosere* along a sand dune area from the beach to the inland ridges.

transfer of technology: the movement of ideas and innovations from region to region or from country to country. One of the most significant movements is from *First World* to *Third World* countries, but even in First World countries technology is often transferred by *transnationals* from countries such as Japan. Japanese work practices, for example, have been readily adopted by European countries.

transhipment is the movement of cargo from one ship to another or from a ship to a different form of transportation.

transhumance is the seasonal movement of animals by people who live in a fixed settlement to pasture elsewhere. The term is usually applied in mountain regions where animals are moved from their lower winter areas to the summer mountain pastures. It can also be applied to other areas of the world where there are seasonal movements, such as the dry grassland regions on the desert fringe.

transition zone: the second zone of the *Burgess* urban land use model; it is also known as the *twilight zone*. In this zone there is to be found the oldest housing in the urban area, which is deteriorating into slum property or being invaded by *light industry*. In *developed countries*, the inhabitants of such areas tend to be of the poorer social groups with some immigrants.

translocation is the movement of soil components in any form (*solution*, suspension) or direction (upward or downward). Soil moisture is usually involved.

transmigration program: the movement of people from one area of a country to another, the best example being seen in Indonesia, where to relieve the population pressure on the main island of Java, people were resettled in the other islands within the country, notably Sumatra, Kalimantan and Irian Jaya. This program has been in operation since the 1930s, and by 1969 the total number of migrants had reached 600,000. In 1969 the Indonesian government initiated a series of five-year plans to handle the movement, and by 1990 the number of families who had moved under these plans was well over one million. The program is not without its critics, as the *indigenous* populations of the host islands do not seem to have been treated well, particularly in the case of tribal land rights. There has also been some concern expressed over the destruction of large areas of the *tropical rain forest* of the area in order to give land to the migrants.

transmittable (communicable) diseases are diseases that can be spread from one person to another. Such diseases can be airborne (spread by droplet infection), fecal-borne (spread by fecal contamination of water, food, utensils, etc.) or spread by direct contact from animals (*vectors*) or humans. Once the method of transmission is known, appropriate, and in many cases standard, preventative measures can hopefully be applied, given the right socioeconomic and political conditions.

transnational (TNC) (multinational): the name for a company that operates in more than one country. Transnationals in the past were typically associated with mineral exploitation or *plantations*, but since the 1950s these firms have increasingly been associated with manufacturing, particularly motor vehicles and electrical goods. Today they cover a wide range of activities including petroleum (BP, Exxon), motor vehicles (Ford, General Motors), electricals/electronics (Philips, Sony, Hitachi), financial services (Barclays), food and hotels (Coca Cola, McDonald's, Holiday Inn). The headquarters of such companies were invariably located within the *First World*, but increasingly transnationals have emerged that are based within the *newly industrialized countries (NICs)*. *Third World* countries have welcomed transnationals, but there are two sides to the argument:

Pros:
- they provide employment and therefore higher living standards
- they may improve the level of expertise of the local workforce
- foreign currency is brought into the country, which improves the country's *balance of payments*
- their presence can lead to a *multiplier effect* and thus an increase in economic activity
- they widen the country's economic base
- they encourage *transfer of technology*

Cons:
- the jobs provided may only require low-level skills
- managerial and supervisory positions may be filled by company employees from the First World
- most or all of the profits may be exported back "home"
- they may cut corners on health and safety or pollution, which they could not do in their "home" country
- they have been known to exert excessive political muscle
- raw materials are often exported and not processed locally; manufactured goods are for export and not the home market
- decisions are made in a foreign country and on a global basis, therefore the company may pull out at any time
- with increased *mechanization*, there is a reduced need for the local workforce.

transpiration is the process by which water is lost from a plant through the stomata (very small pores) in its leaves.

transportation network: the arrangement or pattern of the lines of transportation (road, rail, canal, etc.) that link up a number of places. (See also *network.*)

trench: see *ocean trench.*

trend: a general direction of movement. To describe a trend is to give an overall picture, and not to focus on individual changes.

triangular graph: a graph with three axes, in the form of an equilateral triangle, which allows three separate variables to be plotted. Each axis is scaled from 0 to 100% moving around the triangle in either a clockwise or counterclockwise direction. The advantages of using this type of graph are:

- it allows classification of data in that points that have similar characteristics are located in "clusters" on the graph
- it is the only graph that enables three variables to be plotted.

However, only data that can be divided into three groups and that add up to 100 in percentage form can be used; absolute values cannot be recorded.

It is very useful for example in *soil texture* analysis where % sand, % silt and % clay can be identified, or in the structure of employment in terms of *primary, secondary* and *tertiary.*

Worked example
Sample X (within loam) is:
40% on sand axis
20% on clay axis
40% on silt axis

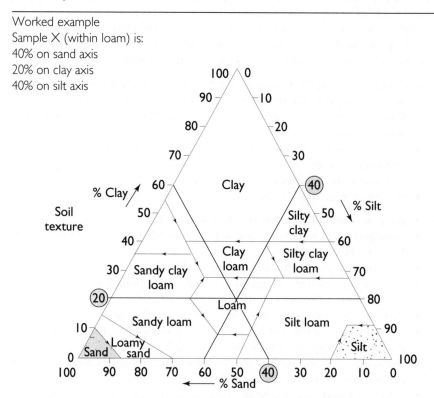

An example of the use of a triangular graph

trickle down: the process associated with *Hirschman's* model of economic development by which economic growth and wealth are spread from the *core* region of a country to its peripheral regions. The expansion of the core increases the demand for food and resources from the *periphery*. The core may also provide the necessary machinery, fertilizers or hybrid seeds to permit the required increases in agricultural productivity. As incomes in the periphery increase, the demand for consumer goods in that area will increase. This may lead to the decentralization of branch plants to the periphery from the core. Governments may seek to encourage this decentralization as the core becomes congested and land prices increase.

trophic levels: the links or stages of the energy transfer model known as a *food chain*. At each trophic level some energy is available as food for the next trophic level. Some of this energy is also lost as excreta, some as the decay of dead organisms, and some as respiration.

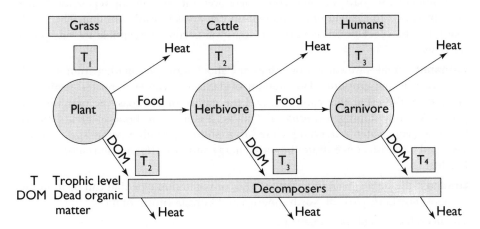

Example of trophic levels in a typical food chain

tropical: a description of those areas of the world that lie between the Tropic of Cancer and the Tropic of Capricorn. The Tropic of Cancer joins the points on the Earth's surface where the midday sun is directly overhead at the June Solstice (June 21). The Tropic of Capricorn joins the points on the Earth's surface where the midday sun is directly overhead at the December Solstice (December 21).

tropical rainforest: see *rainforest*.

tropopause: the narrow layer within the *atmosphere* that marks the boundary between the *troposphere* and the *stratosphere*. It is located at about 16 km above the equator, and 8 km above the poles. The tropopause is characterized by an inversion where the temperature of the limited amount of atmosphere that exists at these heights begins to increase. This inversion marks the upper limit of the weather variations that exist in the atmosphere.

troposphere: the lowest layer of the *atmosphere*, which extends from the ground surface to the *tropopause*. Its main characteristic is that both *pressure* and temperature of the atmosphere within it decrease with height. Over 50% of the total mass of the atmosphere is located within 6 km of the surface of the Earth.

truck farming: the intensive production of vegetables, fruit and flowers for sale. (See *horticulture.*)

truncated spur: a steep wall on the side of a glaciated valley between two tributary valleys. It marks the point where a glacier has removed a former spur that extended into the preglacial river valley. Glaciers are less able to negotiate bends in their course and consequently grind back obstacles in their path.

tsetse fly: the African bloodsucking fly that transmits the disease trypanosomiasis (sleeping sickness), which affects both humans and domestic animals, particularly cattle (where the disease is called nagana). The distribution of cattle in Africa as a consequence is very much dependent on the area of infestation of the fly. The most effective controls have been environmental ones, such as the clearing of woodlands, periodic burning of the bush and destruction of wild game. Other attempts through insecticide, aerial spraying and trapping have not had the same success. Environmentalists have become concerned over the amount of vegetation it has been proposed to clear in order to open up vast areas to cattle farming. The fear is that once the vegetation is cleared the areas will become susceptible to widespread *soil erosion.*

tsunami: a tidal wave caused by submarine shock waves originating from an *earthquake* or volcanic eruption. The amount of damage it causes is related to the distance from the event that caused it, and the nature of the coastline over which it passes. A gently sloping *continental shelf* allows a tsunami to build to great heights, whereas deep water extending close to a shore minimizes wave size. Narrow V-shaped bays also concentrate the wave energy and favor the maximization of wave heights.

tundra is the name given to a climatic and/or vegetation type that can be found in the most northerly parts of North America and Eurasia (north of 65°N).

The main features of the climate are:

- long and bitterly cold winters with temperatures averaging −20°C
- brief mild summers with temperatures rarely being above 5°C
- a large temperature range of over 20°C
- low amounts of precipitation, less than 300 mm, most of which falls as snow

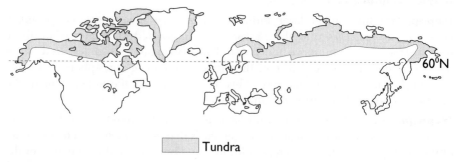

Tundra

Tundra areas

- strong winds blowing the dry powdery snow in blizzards, and creating a high *wind-chill* factor.

Although the winters are severe and the sea regularly freezes, temperatures do not fall as low as more inland areas due to the moderating effect of the sea.

The cold temperatures are due to the short hours of winter daylight in such areas, and although the daylight hours are longer in summer, the angle of inclination of the sun is low. These areas are dominated by high pressure with dry descending air. In summer, some depressions do penetrate, giving some precipitation.

The main features of the vegetation type are:

- dwarf species such as cotton grass, mosses and lichens growing close to the ground often in a cushion-like form. This is an adaptation to the strong winds that blow
- sheltered places may have dwarf willows and stunted birch trees
- the plants have long dormant periods during the cold dark winters, but grow rapidly in the summer when daylight hours are long. "Bloom mats" of anemones, arctic poppies and saxifrages burst into life providing a mass of color
- many plants have small leaves to limit *transpiration*
- the plants are shallow rooted because the soil is permanently frozen at a shallow depth (*permafrost*).

Another feature of tundra areas is the existence of permafrost. The surface layer of this may thaw out in the summer, producing waterlogged conditions in flat areas.

twilight zone: the second zone of the *concentric urban model* produced by Ernest *Burgess*. It lies immediately outside the *Central Business District*, and is an area of old housing that is either deteriorating into slum property or being converted into offices. Light industry may also be invading this area. Its inhabitants tend to be the poorer social groups, with a high concentration of first generation immigrants, and it is also generally associated as being the "red-light" area of the town.

typhoon: the local name for a tropical storm or *hurricane* in the western Pacific Ocean. Typhoons tend to occur off the eastern coasts of China and Japan. (See *hurricane.*)

ubac: a hill or mountain slope that faces north (in the northern hemisphere) and so receives minimum light and warmth. (See also *aspect.*)

ubiquitous: refers to *raw materials* that are evenly distributed and therefore have no influence on industrial location decisions. Water is generally considered to be ubiquitous.

underemployment occurs where people nominally have jobs, but jobs that do not keep them fully occupied. This often results in very low wages, and is particularly prevalent in the agricultural sectors of *Third World* countries.

underground mining: mining at depth is carried out by means of sinking a shaft and then working away from it, usually in a horizontal direction. Problems resulting from underground mining include:

- waste material is also brought to the surface, which has to be stored, usually in large and unsightly tips
- tipping is often done with little regard to the safety of those living below
- this waste can also be carried into streams and rivers, where it causes silting and occasional flooding. Some tips contain poisonous substances that can kill wildlife and contaminate water supplies
- surface subsidies can damage houses in mining areas and create surface ponds and marshes that form in the depressions.

underpopulation occurs when there are too few people in an area to use the resources efficiently. In such circumstances an increase in population will mean a better use of the resources and an increased living standard for all of the population. Canada is often quoted as the best example of an underpopulated country.

UNFPA: see *United Nations Fund for Population Activities.*

UNHCR: see *United Nations High Commissioner for Refugees.*

uniformitarianism: a concept, fundamental in geology, that processes that operate at present also operated in the past, and produced the same results. These processes need not have operated at the same rate, nor at the same intensity. The idea was first put forward by James Hutton in 1795.

Union of Soviet Socialist Republics (USSR) was once the world's largest sovereign state. It consisted of 15 republics of which the Russian Soviet Federated Socialist Republic (RSFSR) was by far the largest. The Soviet Union disintegrated in 1991, following the creation of separate countries from the existing republics. The 15 countries are (in order of population size): Russia, Ukraine, Belorus, Uzbekistan, Kazakhstan, Georgia, Azerbaijan, Lithuania, Moldova, Latvia, Kyrgyzstan, Tajihistan, Armenia, Turkmenistan and Estonia.

United Nations Fund for Population Activities became operational in 1969 with the principal aim of promoting population programs. It concentrates on the following areas:

- family planning – targeted toward both the individual and the family
- programs associated with maternal and child health services in the primary healthcare context
- other forms of education and dissemination of information on family planning
- formulation, implementation and evaluation of population policies.

UNFPA has been guided by two major principles in providing assistance for population activities. Firstly, every nation has the sovereign right to determine its own population policies and program. Secondly, every couple and individual have the basic right to decide freely and responsibly the number and spacing of their children.

United Nations High Commissioner for Refugees: a body of the United Nations that has special responsibility for refugees. It succeeded the League of Nations High Commission and the International Refugee Organization. The UNHCR provides international protection in promoting and safeguarding the rights of refugees with regard to freedom of movement, residence and security against being returned to a country where they may be persecuted as a result of their race, religion, nationality or membership of a political group. The UNHCR can also provide short-term financial assistance to governments that grant asylum to help in the aim of allowing refugees to become self-sufficient members of that community as quickly as possible.

upward spiral: a virtuous circle of growth in which one development leads to a cumulative process. An industry attracts population encouraging service provision, which in turn provides greater employment and spending power within the local area. This is a snowball effect of positive feedback and self-perpetuating growth; the economy is on an upward spiral as illustrated in the process of *cumulative causation* put forward by Gunnar *Myrdal*.

urban area: a built-up area that forms part of a city or town.

urban blight: the run-down or physical deterioration of part of an urban area. This is often linked with areas of social and economic deprivation within the *inner city* and may reflect some uncertainty as to how *derelict land* areas are to be redeveloped. In residential districts the growth of nonresidential activities such as wholesaling or manufacturing can lead to established population moving out to be replaced by very poor urban groups, the most recent migrants, students and other short-term residents. Often these groups do not have the finance or interest to maintain the external fabric of buildings and blight spreads.

urban climate: a localized weather pattern associated with an *urban area*. Towns and cities differ in their climate from the surrounding rural areas in a number of ways:

- temperatures in the cities tend to be warmer – see *urban heat island*
- atmospheric composition – as there is more pollution in cities, there are higher concentrations of gases such as sulfur dioxide and carbon dioxide. There is also more dust and other *particulates* from industrial processes and the exhausts of motor vehicles
- there is a higher incidence of cloud, and therefore lower amounts of sunshine

- the levels of precipitation are greater, with more "rainy" days. Both thunderstorms and the occurrence of hail are more likely
- the frequency of fog is higher, in terms of both its duration and its intensity
- wind speeds are on average lower because of the sheltering effect of buildings. However, within city centers, there is increased turbulence of wind, together with a channeling effect down particular streets.

urban decay: the decline and dereliction of areas of a town or city. Urban decay is a feature of many *inner city* areas of the USA. Economic investment is low, and many of the traditional industries of these areas have closed due to falling demand, lack of space, or the use of new technology. The housing stock is in a poor state of repair. It lacks basic amenities, is overcrowded, and is generally inhabited by people on low incomes who cannot afford improvements. Even where slum clearance has taken place, housing has been replaced by poorly built high-rises, which are themselves becoming unsightly and unsafe. Many stores lie empty, boarded up or protected from crime by metal grills. Graffiti and vandalism are other visible signs of urban decay. (See *derelict land.*)

urban field: a term used to describe the area around an urban center that is functionally linked to that center in terms of the provision of goods and services, administration or employment. More recent terms such as *sphere of influence* or market area are now generally used.

urban fringe: the area on the margins of the urban settlement that lies beyond the city boundary but within the urban sphere of influence. This area includes a variety of residents and activities. Mixed in with the local rural economy, which is still present but in decline, there are *dormitory settlements* housing middle-income *commuters* who work within the city. The fringe is under pressure from developers, which increases land and house prices in the area, and the growth of newer housing areas may lead to a physical segregation, as well as an economic divide, between the *indigenous* population and the new arrivals. Many of these areas are part of the *green belt* and there may be conflicts over the demands for space for further expansion in the urban fringe. With such a concentration of relatively wealthy and mobile population, this area provides economic potential for leisure developments and *out-of-town* malls as well as further residential expansion.

urban function: the use of buildings or land use in an urban area. The broad categories include:

- residential, which may be subdivided by age, quality or form of ownership
- industrial – heavy industry, light manufacturing, industrial parks
- commercial – retailing including services, wholesaling, financial
- offices – medical, legal, agencies (real estate, travel, etc.)
- entertainment – movie theaters, hotels, leisure, clubs, bars
- open land – recreational, parks and gardens, churchyards/cemeteries, derelict/wasteland.

urban heat island: the zone around and above a built-up urban area that has higher temperatures than the surrounding rural areas.

The effect is greatest under calm conditions at night when the urban temperatures

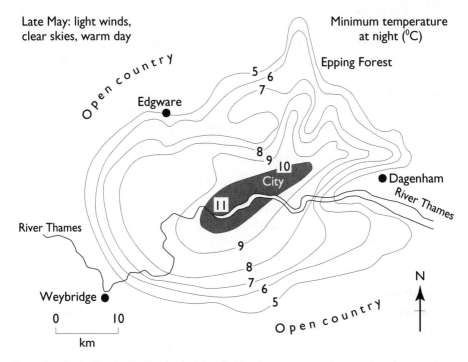

Late May: light winds, clear skies, warm day

Minimum temperature at night (°C)

Example of an urban heat island – London, England

may be 6–8°C warmer. During the day incoming solar radiation is stored in the fabric of the urban structures (brick, concrete, stone and asphalt); this is slowly released at night together with the heat generated in the urban area by industry, space heating systems and vehicle fumes. There is also less cooling by evaporation because surface water is quickly disposed of by the drainage system. *Particulates* and pollution in the atmosphere also help to retain heat. Outgoing radiation may be trapped below the pollution layer. The difference between the urban and rural temperatures reflects the operation of these factors and the scale of the difference is related to some extent to the size of the urban area. There are marked differences between the central area, the inner suburbs, the outer suburbs and the *urban fringe*, which reflect building density and urban function.

urban hydrograph: a graph that shows river *discharge* against time for an area with an urban catchment. The *interception layer* in an urban area consists of buildings, roads and artificial surfaces that reduce *infiltration* and move water into neighboring channels very rapidly through drainpipes and storm drains. Less water falls on natural surfaces such as gardens, parks and open spaces, which encourage infiltration and *throughflow*, and consequently a large proportion of the rainfall reaches the channel quickly to produce a "flashy" *hydrograph*. This has a steep rising limb, a short *lag time*, a high peak discharge and a steep recessional limb. Channels fed by urban catchments are more liable to flooding downstream of the urban area.

urban inequality: the spatial variations that exist in the standards of living urban dwellers enjoy, the range of services to which they have access and the general quality of life they experience. Inequality reflects *social segregation* within cities and is the result of a wide and varied range of factors. Once established, it becomes one of the major concerns for city managers, as it is likely to cause tension and conflict and can even lead to serious social unrest.

urban land use models: attempts by geographers and sociologists to identify and explain variations in spatial patterns that may be similar in many urban areas. These patterns may show differences and similarities in land use and social groupings within cities and they reflect how urban areas have evolved socially and economically through time. The best-known ones include:

- the *concentric zone model* put forward by Ernest *Burgess* and based on his observations of Chicago
- the *sector model* of Homer *Hoyt*, which included some of the concentric patterns seen by Burgess. Hoyt studied over 140 American cities in order to produce the model
- the *Harris-Ullman model*, which was more complex than those based on concentric zones or sectors.

urban primacy: see *primacy*.

urban renewal is the replacement of old structures by new ones and the conversion of space and buildings from one use to another. Renewal involves two processes: *redevelopment*, which involves demolition, and improvement, which seeks to adapt and modernize existing fabric to meet modern needs. Renewal has largely taken place within the *inner cities* of urban areas in both Europe and North America.

urban settlement: the definition of an urban settlement varies from country to country, although the United Nations, for their statistical purposes, define it as a center that has a population of at least 20,000. In Denmark, for example, an urban settlement can have as few as 250 inhabitants, whereas in the USA it has to contain 2500. In some countries the figure is very high; in Japan, for example, an urban area has to contain at least 30,000 people. Geographers have used other criteria in an attempt to distinguish between urban and rural areas. Some of these features include function, occupations, service provision, land use and various social factors.

urban wind: in large cities the wind behaves very differently from the way that it does in surrounding rural areas. The main features of urban winds are:

- urban wind velocities can be up to 30% lower than in rural areas
- periods with no winds at all can be 10–20% higher
- at certain times winds can be extremely strong as high-rise buildings give rise to a funnel effect. It is not unknown for the winds to be so strong as to blow over pedestrians and cause buildings to sway
- there is often small-scale turbulence and eddying (water blown out of ornamental fountains, etc.). This is often the result of the rapid warming of air over concrete and asphalt surfaces leading to the uplift of air, often into strong updrafts, especially alongside tall buildings. There is usually a compensatory downdraft effect on the opposite side of the street.

urbanization is the process in which the proportion of people living in towns and cities increases. Urbanization occurs when rural-urban *migration* is greater than urban to rural, and when *life expectancy* and *natural population increase* are greater in the urban area. Although the process has been going on for centuries, rapid urbanization really only began in the 19th century, and in the *First World*. By 1900, England, for example, was estimated to be already 80% urbanized. In the early 1990s, the UN has estimated that over 70% of the First World population lives in urban areas compared with around 40% in the Third World. Rapid urbanization has occurred and will continue in the *Third World*, to the extent that it is predicted that by 2020 at least 53% of the population will live in cities. This has led to the development of *primate cities* in such areas and the growth of marked differences between the *core* and *peripheral* regions. Some of the problems associated with rapid urbanization include:

- the city is not able to provide enough housing for all of the migrants so they are faced with three alternatives: sleep on the streets; try to rent single rooms; or build their own shelters on land they do not own
- this has led to the growth of *squatter settlements* or *shanty towns*, which have problems of overcrowding, sewage disposal, disease, crime, etc.
- pressure on services such as refuse collection, health provision, education, police and fire services, power supplies and sewage disposal
- transportation systems become overused and the road network is unable to accommodate the increase in vehicular traffic
- the number of jobs available in the city does not match the incoming migration, which leads to vast unemployment. In addition, a large number of people work in the *informal sector* and are therefore classified as *underemployed*.

Studies of Third World cities, however, have shown that the picture of a bleak downward spiral is not always the case. To many migrants, urban life may be very superior. Employment can be found with better wages (see the *Todaro Model*) and many shanty towns are not always areas of deprivation, extreme poverty and disease. Some are very well-organized and not the first destination of recent immigrants to the city.

USSR: see *Union of Soviet Socialist Republics*.

utilities are industries that provide the most basic services such as water and power. Before *privatization* it was generally accepted that utilities were state-run, not only because of their importance to the overall economy, but also because they were operated on such a large scale and were therefore natural *monopolies*.

Valentin's coastal classification: Valentin (1952) classified coasts into two main groups, with an intermediate stage:

- advancing coasts, where marine deposition, or the uplift of the land is dominant. Typical coastal features include *raised beaches, spits, bars, sand dunes* and *saltmarshes*
- *retreating coastlines,* where marine erosion or the submergence of the land is dominant. Typical coastal features include *wave cut platforms, arches, stacks, rias* and *fjords*
- an intermediate stage, which he called "stationary," where deposition = submergence or erosion = uplift.

valley cross-profile: see *cross-profile.*

valley long-profile: see *long-profile.*

value added is the difference between the cost of materials and the price customers are prepared to pay for the finished product. A product with a high value is likely to have been produced and/or designed with great skill.

FORMULA   selling price − bought-in goods and services = value added
$4   − $0.80   = $3.20

The diagram below shows how the $3.20 value added is distributed. It also indicates that value added can be expressed as a percentage:

$$\% \text{ value added} = \frac{\text{value added}}{\text{materials}} \times 100$$

$$= \frac{\$4 - 80¢}{80¢} = 100$$

$$= 400\%$$

$4.00 selling price

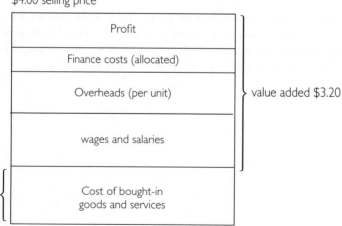

Value added

variable: a measurable characteristic of any person (age, life expectancy, diet), place (GDP, farm size, crop production) or any other thing or quantity.

varve: the annual deposition in lakes near to glacial margins, which consists of a layer of silt lying on top of one of sand. The coarser and lighter colored sand is deposited during spring and summer melt, the darker silts representing the period of decreasing discharge toward the autumn when the finer material will settle. Each combination of light and dark bands thus represents one year's deposition. Counting the varves and measuring their respective thickness gives some indication of the glacial history of the region.

vector: a person, animal or plant that is a carrier of a *transmittable disease*, but is not affected by it. It is a potential source of infection for another organism. For example, with *malaria* the vector is the anopheles mosquito, which hosts and spreads the malarian parasite into the human population.

vegetation succession: see *succession*.

venture capital is risk capital that often provides a significant investment in a small or medium-sized business. The need for it arises when a rapidly growing firm requires more capital, but the firm is not yet ready to raise money on the stock market. In these circumstances, merchant banks might provide the funds themselves, or arrange for others to do so.

vertical integration occurs when two firms that operate in the same industry, but at different stages in the production and supply chain, join together. The integration might come about through a merger or a takeover.

Backward (or upstream) vertical integration means buying out a supplier, e.g. a chocolate manufacturer buying a sugar producer. *Forward (or downstream) vertical integration* means buying out a customer, e.g. a chocolate manufacturer buying up a chain of 7-Elevens.

vicious circle: a series of consequences that act to produce a *downward spiral* of events in which the situation worsens. For example, in regions of poor or inefficient agriculture, low yields means that the farmer has no surplus to sell after the family has been fed; no surplus means that there is no income; a lack of income prevents the purchase of better seeds or fertilizer, which in turn maintains low yield the following year. The situation gets worse because continuous production without soil replenishment through fertilizer will decrease fertility, which will reduce yield even faster. Without *aid*, the farmer's only course of action may be to borrow money at high interest, thus causing further problems. This vicious circle is often referred to as the poverty cycle or trap.

virtuous circle: a series of changes that produce an *upward spiral* as one development encourages growth in another. This leads to a concentration of activity in favored areas such as the *core*. The processes of *cumulative causation* and growth produce *agglomeration* of activity, which attracts other industries, population and services. These in turn make the region more attractive for further services and industries.

visibility: the distance that one can see, which depends on the amount of water droplets, smoke and dust present in the atmosphere. Under international codes, if visibility is reduced below one kilometer, then a *fog* is present, as distinct from *mist* or haze.

visible trade is concerned with the import and export of goods. The difference between visible exports and visible imports is called the *balance of trade*. (See also *balance of payments*.)

volcanic plume: this term can be applied to two phenomena:

- the upwelling of *magma* at a *hotspot* within the *mantle* causing the *crust* immediately above it to weaken and to allow the magma to rise to the surface. Such activity takes place in the middle of the Pacific Plate, and gives rise to the Hawaiian complex of volcanoes
- the cloud of steam, gases and other *tephra* ejected from a volcano during an eruption. The direction of the plume is influenced both by the prevailing wind direction and by the location of the eruption on the volcano.

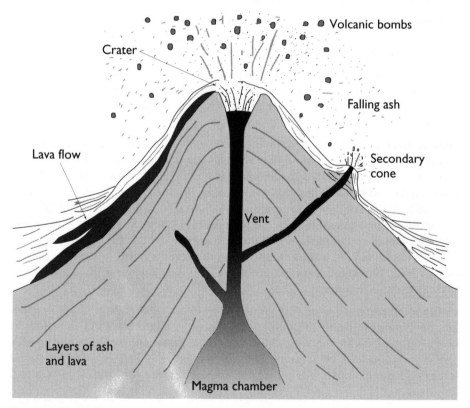

Features of a volcano

volcano: the conical mountain created by *extrusive* materials such as lava and ash that emerge or are ejected from a central vent or crater. There are a large number of volcanoes, and a variety of types of eruption, both of which make the classification of volcanoes difficult. One method of classification is by shape and composition:

- a basic shield – a gentle-sided cone composed of layers of basaltic lava flows, for example, Mauna Loa, Hawaii
- an acid dome – a steep-sided cone composed of viscous acidic lava that quickly solidifies, for example, Mt. Pelée, Martinique
- a composite cone – a volcano built up of alternate layers of lava and ash with steep concave sides. This type of volcano frequently has secondary cones on its slopes. An example is Mt. Etna, Sicily
- ash/cinder cone – composed of fragmental material ejected during a series of eruptions, often with relatively steep sides. An example is Paricutin, Mexico
- caldera – a large circular depression within a much larger volcanic mountain. It is believed that it may be produced by a huge explosion in the magma beneath the volcano, which removed its summit, or by the subsidence of the upper parts of the cones following such an explosion. An example is Crater Lake, Oregon.

von Thunen model: attempts to explain how and why agricultural land use varies with distance from a market. A number of basic assumptions are central to the model:

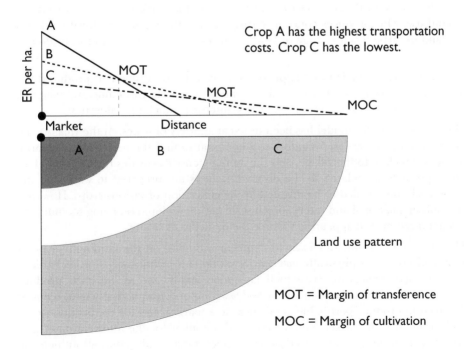

Application of the von Thunen model

- the existence of an isolated state consisting of an *isotropic* plain
- a single central city acting as the only market
- the cost of transportation is proportional to the distance carried
- there is only one form of transportation
- all farmers receive equal prices for the same crop
- the farmers seek to maximize their profits, and they all have the same knowledge of the needs of the market.

Distance from the market is therefore the one variable that would influence the arrangement of land use around the city. At any particular distance the land use that yields the highest net return will be carried out.

A basic principle is that of economic rent, or net profit. Economic rent (ER) is calculated as follows: ER = Price at the market − (production costs + transportation costs).

As transportation costs increase with distance from the market, economic rent decreases with distance from the market. Eventually a point will be reached where ER is zero – this is the margin of cultivation (MOC).

Another basic principle is that of the law of diminishing returns. An increase in production costs through higher inputs does not produce the same increase of return in yield and profit, particularly if greater transportation costs are involved from a more distant location. Hence intensiveness of farming decreases with distance from the market.

As different crops have differing prices, production costs and transportation costs, the rate at which ER decreases with distance from the market varies from land use to land use. This produces concentric rings around the market, each with a different land use type. Each change of land use takes place at a margin of transference (MOT).

Johann Heinrich von Thunen applied his ideas to his own time and produced the following rings of land use: a central city; intensive market gardening and dairying; forestry; arable, becoming progressively less intensive; ranching; wilderness.

The relevance of this model has been tested at a variety of scales. At the village scale in *economically less developed countries*, it is common to find the more intensive farming activities close to the village, and the more extensive activities further away. This does appear to be related to the distance farmers are prepared to walk, and the amount of time needed to be invested in the cultivation of various crops. However, attempts at a national and international scale have foundered on being too much of a generalization, and appearing to force the model to fit reality.

The physical environment clearly has a major influence on the farming type of an area, and no area is physically *isotropic*. Developments in transportation technology have enabled farmers to access their land more easily, and to distribute produce more economically. Developments in *food processing* and *food marketing* have reduced the need to produce perishable goods near to a market, and for the produce to be sold to the nearest market. Von Thunen did not allow for the intervention of governments in farming, with subsidies, guaranteed prices and quotas all influencing production.

It is debatable whether the von Thunen model has much relevance to the location of modern-day farming systems.

voting patterns: the general trends that have taken place in elections at a local, regional or national scale reflecting the changes that have occurred in the political environment.

vulcanicity refers to all of the processes by which gases, liquids and solids are either injected into the Earth's crust or ejected on to the surface. It therefore refers to *intrusive* features such as *dykes* and *sills*, and also *extrusive* features such as a *volcano*. Vulcanicity is closely related to *orogenesis*, and usually occurs when mountain building is in progress or when severe crustal movement takes place.

wadi: a steep-sided ravine in desert and semidesert areas, usually streamless, but sometimes containing a torrent (*flash flood*) for a short period after heavy rain. Also known as an arroyo in the Americas. The infrequency of flash floods compared to the great numbers of wadis in some areas suggests that they were created at a time when storms were more frequent and severe.

Walker cell: an east-west tropical *atmospheric circulation* in the South Pacific first recognized by Gilbert Walker in 1923. This circulation covers a broad area and is associated with strong convectional activity.

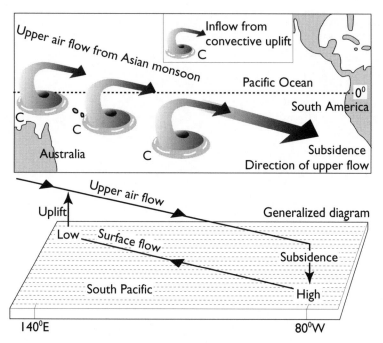

The circulation of the Walker cell

warm front: the leading edge of a mass of warmer air usually seen in a *depression*, where it is the leading edge of the *warm sector*. Such fronts are associated with thickening cloud and, almost inevitably, some form of *precipitation*. A warm front is not as steep as its cold counterpart, having a gradient of between 1 in 100 and 1 in 250.

warm sector: the area between the two fronts of a *depression* where the highest temperatures are recorded. Cloud tends to be low (*stratus*) or broken, with *precipitation*, if any, in the form of light rainfall or drizzle. The low cloud of the warm sector often drifts onto upland areas where it forms hill *fog*. As the *cold front* catches the warm, the size of the warm sector decreases until the warm air is finally pushed aloft as *occlusion* occurs.

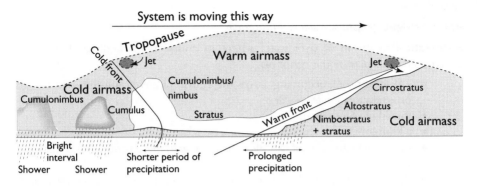

System is moving this way

Warm sector

waste disposal: the disposing of the by-products or rubbish (unwanted materials) produced by man in his use of the Earth's resources and the manufacture of goods. Disposing of waste is an increasingly difficult task, the nature and quantity of modern waste making it so much harder to cope with. Some waste is hazardous and many environmental organizations have increasingly campaigned for greater safeguards in the disposal of such materials. The places for the dumping of modern waste include the following:

- *landfill sites*, particularly for domestic rubbish and chemical waste
- rivers, for low-level chemical waste and the treated water from sewage works
- the sea, although organizations such as *Greenpeace* have been actively campaigning for this to stop
- the *Third World*, for the waste products from the First. Some countries have been willing to take hazardous waste from the *First World* in return for large payments, although the lower safety standards of such countries has given rise to some serious concern.

waterborne diseases are those that are essentially spread by insects and other very small animals that breed or are found in water. The major types, all of which are *endemic* to areas of the *Third World*, are:

- *malaria*, spread by the female anopheles mosquito, which breeds in stagnant water
- yellow fever, which is also spread by a mosquito, although a different species
- *bilharziasis*, which occurs when a parasitic worm living inside freshwater snails invades the body
- river blindness, caused by the blackfly, which again infects the body with a parasitic worm. The fly lives in fast-flowing waters that are highly oxygenated
- guinea worm, which invades the body after water is drunk containing the larvae of the worm.

Typhoid and cholera are also caught by drinking contaminated water. Good hygiene and the availability of a safe water supply are essential for the control of these diseases.

water budget graph: see *soil moisture graph.*

waterfall: a steep fall of river water where its course is markedly and suddenly interrupted. This may be the result of:

- a resistant rock occurring across the course of a river and interrupting its progress to a *graded profile*
- the edge of a plateau
- a fault-line scarp
- the overdeepening of a valley through *glaciation*, producing a hanging valley
- the *rejuvenation* of the area, giving the river renewed erosional power, which in turn gives rise to *knickpoints* where profiles intersect.

waterlogged describes the state of the soil when all the pore spaces below a certain level are full of water. If this extends up to the surface no water can enter the soil, which results in surface flooding. The region below the *water table* is thus waterlogged.

water management is needed to maintain the quality and quantity of the water supply. Countries in the *First World* have the financial resources and the technology to develop large schemes but in the *Third World* there are problems in many areas in developing and maintaining a good water supply. Many water management schemes in such areas have to be multipurpose in order to justify the cost. Modern schemes also attempt to manage the whole of a drainage basin, even if, as is the case with the Tennessee Valley, that basin covers several states. Water management there consists of collection, distribution and the prevention of flooding.

water quality: in the 1980s the United Nations through various agencies aimed to make clean water and sanitation available to everyone by the end of the decade. Although great strides were made, the biggest obstacles to success have been the limited financial and technological resources in the *Third World* and the rapidly growing populations of those countries, particularly the expansion of squatter settlements within urban areas. Despite attempts to clean up water supplies, over 20 million people still die every year from *waterborne diseases* such as *bilharziasis*, guinea worm, cholera and diarrhoea. Millions more are affected by the insects that breed in freshwater and cause *malaria* and river blindness.

watershed: see *divide.*

water table: in a rock when all the pore spaces are full, there is a zone created that is said to be saturated (*waterlogged*); the upper boundary of this layer is known as the water table. The water table will move up and down, depending on the supply of water from above and the amount of evaporation from the rock and soil.

wave cut platform: the retreat of a cliff leaves behind a gently sloping area that cuts across the rocks of the coastline regardless of type and structure. As the platform grows, incoming waves break further out to sea and have to travel across a wider area, which dissipates their energy and reduces the impact upon the cliffline. Some wave cut platforms are now found at a higher level as *raised beaches*, indicating that there has been a negative change in sea level resulting from a fall in sea level or a rise in the land surface.

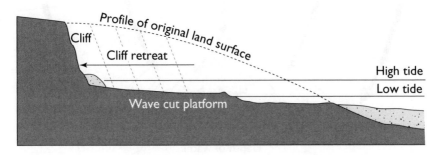

Development of a wave cut platform

wave formation: waves are caused by the frictional drag exerted by winds as they pass over a water surface. When a wind passes over a water surface it tends to generate waves that move forward in the direction in which the wind is blowing, but there is little or no forward movement of the water itself. The size and energy of the waves produced depend upon three factors:

- the velocity of the wind – the stronger the wind, the larger the wave
- the period of time during which the wind has blown – the longer the wind blows, the larger the wave
- the *fetch.*

wave refraction is the tendency for waves to become parallel to the line of a coastline. It is due to the fact that a wave approaching a shore loses energy as the depth of water decreases. This can be illustrated by the example of waves approaching an irregularly shaped coastline, such as a headland and bays. The waves approaching the headland find that the water shallows more quickly and the movement here is slowed down. However, the waves still in the deep water of the bays are unaffected and move more rapidly until they reach shallow water further in. The line of the wave approaching the coastline therefore begins to reflect the shape of the submarine contours. Another consequence is that the erosional effect of the waves is concentrated on the headland. This may also cause a slight local rise in sea level at the headland, resulting in a longshore current from the headland to the bay.

waves are elliptical or circular movements of water near the surface of the sea or any other body of water. They can be described by a range of terminology:

- crest and trough are the highest and lowest points respectively
- wave height is the vertical distance between the crest and the trough
- wavelength is the horizontal distance between two crests
- wave period is the time taken for a wave to travel through one wavelength.

Wave motion decreases rapidly with depth, and there is little movement at a depth greater than half of one wavelength. (See also *constructive waves, destructive waves,* and *clapotis.*)

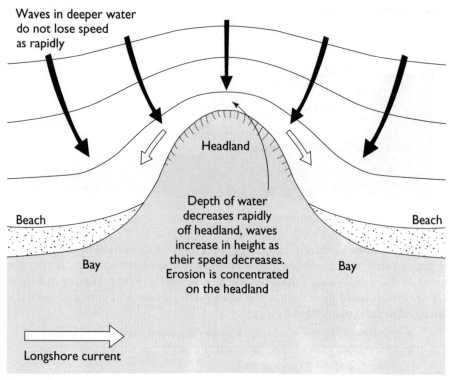

Wave refraction

weather: the state of the atmosphere at a particular point at a specific time. The weather may be described in terms of temperature, precipitation, wind speed, wind direction, cloud type, humidity and visibility.

weather forecasting: the prediction of what the weather of a particular location is going to be. To do this a great deal of precise information is needed, and it is collected through observations from satellites, airplanes, ships, weather balloons and a network of recording stations. With the help of computers and past records meteorologists are able to produce charts and forecasts that are constantly updated.

weathering: the disintegration and decomposition of rocks in situ by the combined actions of the weather, plants and animals. (See also *chemical weathering* and *physical weathering*.)

Weber, Alfred: produced a model to explain and predict industrial location. A number of basic assumptions are central to the model:

1 the existence of an isolated state consisting of an *isotropic* plain upon which movement is equally easy in all directions
2 some raw materials are *localized*, i.e. they are not evenly distributed
3 some raw materials are *ubiquitous*, i.e. they are evenly distributed
4 each of these raw materials may either be gross (they lose weight during manufacture) or remain pure (all of their weight is included in the finished product)

5 transportation costs are dependent on the weight carried and the distance traveled

6 the size and location of the markets are fixed points, and labor costs are the same at all points across the plain

7 perfect competition exists across the plain. In other words no single manufacturer can influence prices of either the raw materials or the product.

Weber stated that an industrialist would establish an industry at the *least cost location*. He devised the *material index* to assist in the identification of such a location. (See also *agglomeration, isodapane, isotim*.)

Weber's model can be criticized in a number of ways:

- his assumptions are unrealistic. Transportation costs are not directly proportional to distance, but tend to increase in a stepped pattern. They vary according to different modes of transportation, and factors other than weight and distance are also important, such as fragility, perishability and bulk
- the model does not take into account the role of governments in influencing the location of industry or targeted private investment
- markets are rarely at one point because demand covers a wider area. It has been suggested that the least cost location is the wrong tenet upon which to base the model – it should be the location of maximum profit
- some people believe that industrialists are not profit maximizers – they are more interested in *satisficing*
- the role of *transnational* corporations is a major factor. Industries are increasingly aimed at a global market rather than a national market, and the way an individual factory fits into the overall strategy of the corporation is more important.

Wegener, Alfred Lothar: the German meteorologist and geophysicist who formulated the first complete statement of the *continental drift* hypothesis. In 1915 he published his work on the origins of the continents and oceans, suggesting that in the late Paleozoic era all the present-day continents had formed a single land mass, which had subsequently broken apart. The name *Pangaea* was given to this supercontinent. Wegener's theories did not find widespread acceptance, particularly as his suggestions as to the driving forces behind the continents' movements seemed implausible. By 1930 his theories had been rejected by most geologists, and sank into obscurity, only to be resurrected as part of the theories of *plate tectonics* in the 1960s.

welfare indicators are used to measure the state of well-being of an area of a country, or for a country as a whole. They give an indication of the *quality of life* of that area or country. They may include measures of:

- health – for example, *infant mortality* rates, numbers of people per doctor, *life expectancy*, nature of diets
- incomes – for example, wage levels, unemployment rates, number of households with retired members, numbers of children in receipt of free school meals

- provision of basic services – for example, households lacking or sharing a bath or toilet
- social indicators – for example, crime rates, levels of vandalism.

wells are holes dug into the ground for the purpose of obtaining underground supplies of water. They should be sunk into an *aquifer*, and go some distance below the level of the *water table*.

Much of the aid donated to *economically less developed countries* is being used for the purpose of drilling tube wells. These have concrete walls and are covered, and the water is extracted using a modern pump.

wetlands are areas where the soil is frequently or permanently waterlogged, with the *water table* being at or near to the ground surface. Their existence has steadily declined in recent years as they are drained for either arable or pastoral farming, or for the exploitation of their peat supplies. They provide a distinct habitat and are the home of a variety of rare plant, animal, bird and insect species.

The establishment of *environmentally sensitive areas* is seen as a possible way to achieve both the protection and environmental improvement of the wetlands.

wetted perimeter is the total length of a river's bed and banks that are in contact with water when viewed as a cross-section.

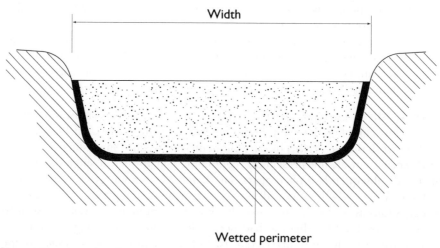

River width v. wetted perimeter

white collar workers: people who work as professional, managerial or clerical staff.

white goods is a collective term for household kitchen appliances such as refrigerators, stoves and washing machines, all of which were traditionally finished in white enamel.

WHO: see *World Health Organization*.

wholesaling: the process of buying large quantities of goods from suppliers and selling on in smaller volumes to retailers or business users. This role is often looked down on because the wholesaler's profit margin appears to make goods more expensive for the consumer. In reality, wholesalers reduce the number of journeys that manufacturers would have to undertake if they wished to service each outlet at which their goods were sold. Wholesalers therefore reduce manufacturers' costs.

wilderness: large areas that have been set aside for protection. These are regions that are remote and where the ecology of the area has never been seriously affected by human interference. They also lack a deliberate policy of management. Scientific reasons for establishing wilderness include:

- the need to maintain the gene pool of wild organisms to make sure that genetic variety is maintained
- animal communities should be kept to their natural environments to allow research into their ecology
- the need for a pure natural system as a yardstick against which managed or mismanaged ecosystems can be observed and measured.

There are strict controls within wilderness areas. Hunting and trapping, commercial development and motorized transportation are prohibited or seriously restricted. Alaska has 35 such areas, which total over 300,000 square kilometers.

wildlife refuge: an area largely free from the impacts of human settlement and development where animals, birds, insects and plants can exist in harmony with their natural environment. They have a high conservation value, and for this reason people wish to visit them. The management strategy for such an area must be based on a low carrying capacity in that human impact must be kept to a minimum.

wind chill: the accentuation of cold temperatures when accompanied by high wind speeds. On a cold but calm day, warm human skin heats the air next to it, which in turn passes heat to the next layer and so on until heat is taken away from the body. This is an inefficient heat transfer mechanism as the conductivity of air is low. Therefore, calm air feels relatively warm. A wind, however, pulls heated air away from the surface of the skin and replaces it with colder air. The faster the air moves the more heat is taken from the body, making the cold air feel much colder than the reading on a thermometer. For example, an air temperature of 0°C with a wind speed of 6 m/sec. has a wind chill factor of −10°C.

wind deposition occurs when the moving air no longer has the ability to carry the amount of sand and other particles that it was transporting. A number of *aeolian* landforms are created, for example, *dunes, barchans* and *seifs*.

wind erosion is the removal of sand and other particles from the ground surface and the subsequent use of those materials to wear away rocks in other areas by the force of moving air. The main processes involved are *deflation, saltation* and *abrasion*. The *aeolian* landforms produced include pedestal rocks, *yardang* and *zeugen*.

A pedestal rock is a mushroomed-shaped rock that has been eroded more just above its base than at higher levels. This is because the sand grains carried by the wind are rarely lifted more than one meter above the ground surface, and so perform little erosion above this height. The shape may be accentuated if there is a less resistant layer of rock lying beneath a resistant layer.

wind formation: winds result from the differences that exist in air pressure between different places. These differences occur because of variations in temperatures between places. When the air temperature of a place increases, the air in that area expands and rises, thus reducing air pressure. Conversely, when temperatures fall the air becomes more dense and air pressure increases. The gradual change of pressure between different areas is called the pressure gradient, and it gives rise to a movement of air from the area of relatively high pressure to the area of relatively low pressure. This movement of air is wind.

The strength and direction of the wind is influenced by the *pressure gradient*, and by the *Coriolis Force*.

wind power is one of the few *renewable* sources of energy being developed commercially on a large scale. For it to be efficient, a wind turbine should be located at a point where wind is both regular and strong. Suitable sites include hilltops and coastal areas. It is common to find many turbines grouped together to create a "wind farm." The initial cost of setting up a wind farm is expensive, but the power it produces is relatively cheap. Wind farms may also attract small businesses to areas that otherwise would be lacking in job opportunities.

However, wind power could never provide large quantities of electricity. Moreover, there are already concerns about the unsightliness of the wind farms that currently exist.

wind transport refers to the movement of sand and other particles through the air. It is greatest where winds are strong, turbulent, blow from a constant direction and blow for a long period of time. It is more likely to take place when the ground is dry, unconsolidated, and sparsely vegetated. The particles on the ground are therefore loose and available to be set in motion.

Wind moves material by three main processes:

- suspension
- *saltation*
- surface creep – the rolling forward movement of larger particles resulting from the impact of falling sand grains. (See *saltation*.)

WMO: see *World Meteorological Organization*.

workers' cooperative is a business owned and controlled by those who work in it. Large-scale cooperatives of this kind were attempted in the 1970s, but they failed because, despite the additional commitment of the workforce, the businesses were simply not viable. They also faced the problem of management, as it was often necessary to bring in outsiders to perform vital management functions, such as accountants and marketing experts. This runs counter to the cooperative ideal and was the source of much tension, leading to considerable inefficiency. This is less apparent in the small-scale cooperatives set up in the 1980s, which were numerous and often successful.

World Bank, or the International Bank for Reconstruction and Development, was established in 1947 to provide aid to *developing* countries in the form of loans and technical assistance. Originally the Bank supplied loans for capital projects, such as those concerned with improving a country's *infrastructure*, but from 1980 onwards it was allowed to offer help in the case of *balance of payments* difficulties, subject to conditions.

World Health Organization (WHO): a specialized agency of the United Nations, based in Geneva and established in 1948 to further international cooperation for improved health conditions. The work of WHO includes:

- providing a central clearing house for information and research on such features as vaccines, cancer research, nutrition, drug addiction and nuclear radiation hazards
- sponsoring measures for the control of *epidemic* and endemic diseases by promoting mass campaigns involving mass vaccination programs, instruction on the use of antibiotics and *insecticides*, assistance in providing pure water supplies and sanitation systems, and health education for rural populations. The eradication of smallpox by 1980, for example, was due largely to the efforts of WHO
- encouraging efforts to strengthen and expand public health administrations of member nations.

World Meteorological Organization (WMO): a specialized agency of the United Nations, founded in 1951, and created to promote the establishment of a worldwide meteorological observation system, the standardization and international exchange of observations and the development of national meteorological services in *Third World* countries. Its administrative headquarters are in Geneva.

World Park: a recommendation from the second World Conference on National Parks (1972) whereby Antarctica and its surrounding seas should be designated as the first World Park, where it would be a sanctuary for wildlife and a continuing outdoor laboratory. Although there is support for such a designation, nothing had happened by the early 1990s.

World Wildlife Fund (World Wide Fund for Nature) (WWF): an organization, founded in 1961, whose aims are to protect endangered species and to tackle all environmental problems that threaten any form of life. Among its recent concerns have been the tropical rainforests, marine conservation, the preservation of wetlands and pollution within the *European Union*. Its headquarters are in Switzerland.

xerophytic describes the vegetation that is adapted to living in dry conditions. Drought-resistant plants have some of the following features:

- thick, corky barks to cut down transpiration
- bulbous trunks made of spongy wood that stores water
- widespread branching root systems to gather water from the widest possible area
- large proportion of the *biomass* is below the surface
- small leaves to cut down the surface area through which water can be lost
- sunken stomata that only open at night to allow respiration, and to take in carbon dioxide for the subsequent production of organic matter during the day
- waxy surfaces on the leaves to help reduce water loss
- reduced surface area by having spines instead of leaves
- folded leaves with the stomata inside
- deep penetrating roots that seek water at depth (tap roots).

xerosere is a name applied to a plant *succession* that develops under markedly dry conditions. It may comprise either a rocky sere (*lithosere*) or one formed on sand (*psammosere*).

yardangs are typical of desert and semidesert areas and are extensive ridges of rock separated from each other by grooves or troughs. They are aligned in the direction of the *prevailing wind*. Their height can vary from one meter up to well over fifty meters.

yield in agricultural terms is the amount produced, usually expressed in terms of the area involved, e.g. tons per hectare.

yuppie: a 1980s label describing a supposedly new class of young upwardly-mobile professionals devoted to smart but faddish lifestyles. The provision of dwellings for such people has had a major effect on the inner areas of major cities, particularly in the process of *gentrification* within the *inner city*.

zero growth is a situation in *demographic* terms where the population reaches replacement rate and there is no growth. It is predicted that in the early years of the 21st century this will be the norm for the USA and Europe. This situation can have its drawbacks:

- an aging population, and the cost of providing for it. For example, 20% of Japan's population will be over 65 by 2007
- labor shortages in many key industries
- the need for those in work to pay for the elderly, particularly in terms of pension provision.

zeugen: in desert and semidesert areas these are tabular masses of resistant rock, standing out from softer underlying rocks. They are produced by differential erosion through the scouring effect of sand-laden winds.

Zipf, George Kingsley: popularized the *rank-size rule* in books published in 1941 and 1949.

zonal soil: a classification of soils covering those types that show the maximum effects of climate and vegetation on soil formation. Zonal soils are mature soils, in that they have had time to develop distinctive profiles with clear horizons. Examples of zonal soils are *tundra* soils, *podsols, brown earths* and *chernozems.*